DESIGN MIND FOR DATA VISUALIZATION

J.STORM

DESIGN MIND

FOR

DATA VISUALIZATION

LEARN TO USE INFORMATION AND GRAPHIC DESIGN PRINCIPLES TO PRODUCE ENGAGING DATA STORIES YOUR AUDIENCE CAN'T IGNORE

6X9 PH

Cover Design: Belma @planb.dsgn

Published by 6X9 PUBLISHING HOUSE LLC
www.6x9ph.com

ISBN 979-8-9869394-0-7

For you, the reader

CONTENTS

PART TWO
ELEMENTS AND PRINCIPLES OF DESIGN

PART THREE
APPLY YOUR DESIGN MIND TO DATA VISUALIZATION

INTRODUCTION

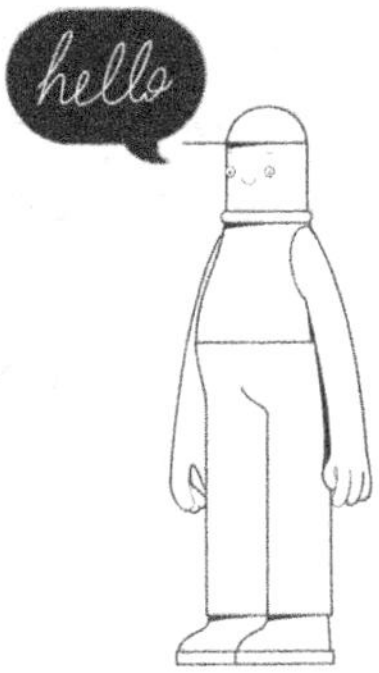

Design Mind for Data Visualization is a book that focuses on explaining how a visual is perceived before going into the specifics of its creation. It teaches the reader about visual perception, information design, and the fundamental elements and principles that must be applied to create a powerful data visualization. If you have come across this book, here are three reasons you should stick around to read it:

1. You are overwhelmed by the fancy designs of your peers and have zero knowledge of graphic design.
2. You need some inspiration to take your work to the next level.
3. You are struggling to hook your audience's attention to your graphs and are in dire need of some help.

Well, you are in luck because this book can help you with all three problems. Designing a visualization is not rocket science, and anyone can lay out a graph or chart without trying too hard. But does your audience connect with your work? Are you proud of your finished design? If you are looking to design exceptional visualizations, this book will help you with your journey. It aims to help someone that wants to make their vizzes aesthetic, meaningful, inclusive, and emotionally engaging. It also breaks down the entire process of creating, designing, and perceiving a visualization so you can design sensibly and sensitively.

The first part of this book, 'Eyes, Brains, and Hearts, Oh My' focuses on the science of perception. How do people actually see and take in a visual? It breaks down the process of seeing, perceiving, and retaining information when someone lays their eyes on your graphic. It delves deep into how that process directly affects design decisions. The section mainly supports the idea that a thorough understanding of the human perception of design is crucial and supports the process of creating inclusive, efficient, and informative data visualizations.

The second part of this book, 'Elements and Principles of Design,' starts with graphic design basics to take you from design-aloof to design-savvy. It focuses on the building blocks that one must know to create a challenging and visually engaging design. Once you are introduced to the basics such as line, shape, balance, contrast, layout, etc., you will learn to be visually aware and use these design principles in all of your future work. Applying these elements and principles thoughtfully and strategically can create a great visual that speaks volumes.

The third and most important part of the book, 'Apply Your Design Mind to Data Visualization,' focuses on applying all the knowledge from parts one and two to your future designs. It goes into explicit detail about how you can create a successful visualization utilizing the principles fundamental to information design. In part three, I talk about how to create a process for yourself and build out your own dataviz style guide to maintain consistency throughout your projects. I also dive into creating a design process that will help speed up your work and avoid rework. This part of the book also focuses on accessibility, diversity, and inclusion in design. Since the core principle of design is to know your audience and cater to their needs and wants, the book urges the reader to consider all empathetic possibilities before delving into the design phase.

As a data author, displaying your data story to express maximum information with minimum confusion is crucial. Design is visual communication, and your viz should attempt to inform, influence, or entertain your audience without you having to speak for it. It will be harder to do so without considering your audience's perspective during the design phase. This work comes with great responsibility. Only when you understand the who and their why can you design impactful and meaningful visualizations. You must know who you are designing for before you can work on what you are designing and how you will design it. Each step is explained in considerate detail because becoming fluent in visual vocabulary and skills must go hand-in-hand with understanding visual disabilities, global audiences, human perception, and emotion.

Finally, the fourth part of the book, 'Moving Forward,' offers ways to get your creative juices flowing and maintain that flow. Technologies, styles, demands, and trends tend to change rapidly, and if you want to be a successful data author, you must take it on with dedication, passion, and love in your heart.

Design education is a lifelong process that requires you to constantly study, practice, and challenge the inner workings of the human mind. With this book as your guide, you will open new doors to your creativity and insight into a field that will keep you on your toes with delight.

This book is written with much research, insight, and dedication to help bored and unchallenged proficient data analysts who want to expand their skills creatively. It will also help guide data visualization enthusiasts looking to build their professional portfolios or those who need simple and practical design solutions to improve their presentations and visualizations for work. This book is here to guide people who are confused about where to start with their designs. It will give you an excellent understanding of fundamental design principles to build up your mental design toolkit so you can apply all of this information to your data visualizations. It shows you how to consider size, shape, and contrast to grab attention, compose a layout in a way that doesn't overwhelm, design with empathy, and much more. The book will give you a new perspective on how to look at the world and how vast the audience could be, especially online. It teaches you to be thoughtful of your designs and create memorable experiences to incite positive emotions in audiences. Through your self-discovery of design, you begin a journey to realize your untapped creative powers to make data visualizations that don't just capture the mind but also the heart.

BONUS GIFT

Just to let you know before you start reading, in order to reduce printing costs, the images in this book are black and white.

For those color lovers out there, I have made a special gift just for you. Because you're awesome and I want to thank you so much for purchasing this book.

Go to the link below or **scan the QR code** to access a PDF copy of this WHOLE BOOK with COLOR images. For free.

dmdv.colorbonus.6x9ph.com/bonus

PART ONE
EYES, BRAINS, AND HEARTS OH MY

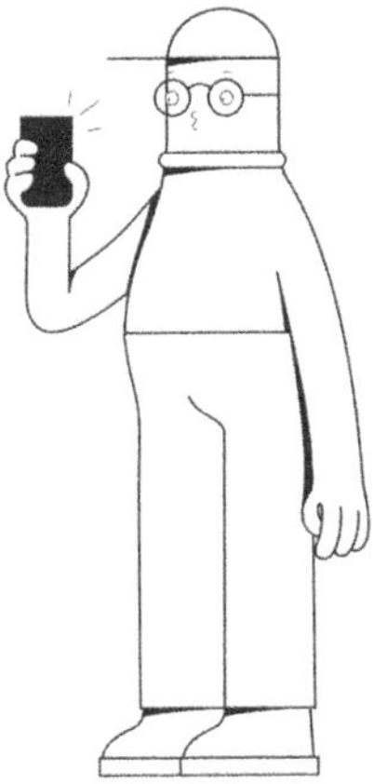

"The greatest value of a picture is when it forces us to notice what we never expected to see."

JOHN TUKEY

CHAPTER 1
ARE VISUALS THAT IMPORTANT?

IN A WORLD LIKE OURS, everything is based on convenience. People only peruse a website if it's convenient. They only go to parties, attend meetings, and stick to plans if they are fitting. And, they only retain information if it is presented conveniently. This isn't our fault alone, no. It's because the human brain is lazy and prone to shortcuts.

The (lazy) human brain is designed to accept visuals over detailed text. Therefore, everyone has no choice but to be attracted to visuals and immersed in a visual experience. This is because vision is the most dominant sense for most of the population with intact optical systems. More than one million nerve fibers are present in our eyes, accepting, translating, and analyzing this visual data at an extremely high speed. Furthermore, the human brain has a considerable capacity to retain picture information faster than text and with very few errors. According to science, the human brain has the potential to process an image 60,000 times more quickly than text. This goes to prove that while it takes only a few minutes to read and understand the information provided in a 500-word text, a visual can do it in only a few seconds with a more substantial, more lasting impact.

THE POTENTIAL OF A GOOD VISUAL

So, what are visuals? Are they that important? What are their origins, and how did we come to identify them as a language for communication? These very valid questions surface in our minds when we think of any form of art. But since Design is a field that deals with a significant part of visual communication, designers are the ones who often get lost in this sea of questions.

We think of a visual/picture/infographic as a two-dimensional surface often marked with colors and shapes. But a picture is much more than that. An image is a form of communication that tends to transcend all languages. It reflects the creator's ideas and concepts in a way that enhances the connection with the audience and evokes a powerful visual experience. When it comes to design, there is a careful, playful, and deliberate use of color, lines, and shapes to transfer a message to the viewer. It is assumed that a particular concept will be transmitted when the viewers see those sets of graphics, and they will proceed to analyze that information in an orderly sequence.

More than 91% of the human population leans towards visuals more than text. This is why data is described everywhere you go in easy-to-understand, vibrant, and interactive graphics, making it easier for the viewer to comprehend while saving you a bulk of time. In an increasingly information-saturated world, efficient and accurate information is crucial. Visuals are an excellent way of communication. Every time a designer comes forward with a visual, they try to convey a message to the audience in a simple way that everyone understands. Whether it is an artwork or an infographic, a designer will also try to convey the most information they can in the best possible way. Designers also tend to capture abstract ideas that are too difficult or vague to explain in words and convey them through a visual. Visuals help us quickly, effectively, and persuasively share profound, meaningful messages.

A visual can become a great motivator and inspire people to think a certain way. Graphics tend to subconsciously affect our minds whether we remember seeing the visual or not. They hold great power and can completely alter a person's thinking. This is why the field of Design is a meeting place between branches of both art and science. Everything from typography, color, shapes, layout, and size must be intentional because they affect how you perceive something. Just by altering the size of a text or positioning shapes differently, the meaning of your visual can drastically change. You can have a significant impact on the world with just a graphic or image. That is the amount of power a visual holds.

Visuals can have a particular effect on people. They tend to make people feel a certain way and the feeling can vary for each person based on their unique interpretations and experiences. That is why for each person, design can mean something different. We tend to look at beautiful things and term them attractive because they make us feel a certain way. People long to acquire beautiful things that make them feel special to constantly surround themselves with that feeling. Visuals in design can invoke various emotions, from calm to excitement to curiosity.

The five intertwining rings of the Olympics symbol is an excellent example of a very simple visual that is globally recognized. It stirs excitement for viewers and heated determination in athletes who dedicate years of their lives for a chance to stand on the podium. You'll often see winners with Olympic ring tattoos to mark the memory.

THE IMPORTANCE OF DESIGN

A well-made visual has the potential to make our lives better and easier by helping us make fast, effective decisions and redirect bundles of time to other important tasks. Design is found everywhere, from road signs to smartwatches and of course, graphs and tables. A visual transcends language boundaries and unites the whole world under imagery. For example, the tedious task of reading, translating, and analyzing a manual can be made extremely simple with a visual aid.

• • •

Think back to the last time you bought some IKEA furniture. How long did it take for you to assemble? Hopefully, it was not too long, right? Even if making that shelf or desk took longer than expected, now imagine completing it without any visual aids!

Imagery has had a significant influence on humanity since the very start. Before the human mind could come up with language, it came up with visuals to effectively communicate with others. From the primitive man and his cave drawings to the visuals we use today, each has a particular meaning, intention, and reason behind it. History is full of proof of how visuals have the potential to bring about change and invoke a wide range of feelings in a person. For example, during the Romantic period, artists most influenced by war and rampage took to visuals to express their trauma, misery, and affliction. These visuals not only represented the thoughts and feelings of the artist but also helped viewers resonate with those emotions and, in turn, share their similar feelings.

Another great example is Picasso's famous painting "Guernica" which he painted after the Spanish Civil War in 1937. This masterpiece carefully and sincerely expressed all atrocities, trauma, and collateral damage that war left behind on the people of Guernica. The huge monochrome painting smartly depicts how a world full of color is left drained as a terrible aftermath of war. The composition highlights the presence of gorged animals, women screaming, an eye, flames, and a dismembered soldier. All these elements not only portray that war affects everyone from humans to animals but also recognize and speak to the suffering of each individual. This is an excellent example of how an influential, meaningful visual can help speak to people on an emotional level.

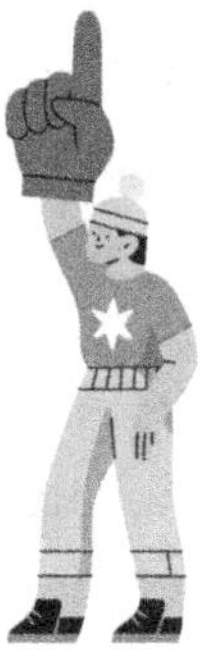

Avid supporter

Whether a visual is used for selling a commercial product, promoting a non-profit company, or communicating news program statistics, a good design has the power to unite people under one cause.

Furthermore, design can directly impact an individual's life since all of our surroundings are built on innovation. In such a world, it becomes impossible to live independently from design. A tool that can impact decisions and trigger someone's emotions and feelings is a powerful one and cannot be ignored. You must use it to your advantage.

GOOD DESIGN IN DATA VISUALIZATION

In data visualization, a good design can turn lengthy, complex, and confusing data into an engaging visual. Data becomes not only basic information but also digestible and easy to understand. With a skill like this, your message can be received by a wide audience. You would expect that your design would translate through to everyone, from the CEO to a casual internet scroll-er.

With all the mention of a "good design," one starts to wonder what constitutes a good design. Is it the way you design a visual? Or the idea it communicates? Or how well the idea is perceived by the audience? Is it one, or is it a mix of the above? Unfortunately, there is no hard and fast rule of what makes a well-designed visual, but there are some ways of making sure your design is clear, thorough, and goes through to your audience. These may include:

- Forming visuals that are easy to identify
- Creating a layout that is straightforward and free of clutter
- Organizing your content so that it is swiftly comprehended by the viewer with minimal effort
- Acquiring maximum impact with precise use of fonts, shapes, and colors
- Incorporating contrast, hierarchy, and balance to draw the attention of the viewer to the most important attributes

As humans, we tend to take our senses for granted. Interpreting, evaluating, and taking in the world around us through our eyes, ears, and hands is so natural that we often forget to pay attention to it all. As a result, we go through more than a hundred different types of visuals each day without genuinely paying attention to them. The ability of a human brain to absorb and process visual information is called visual perception. Therefore, we must understand the process behind how we perceive visuals so that we can confidently make design decisions for ourselves.

Perception is a cognitive process, and some of the key players in visual perception are the eyes and the brain. The brain interprets data that is collected by the sensory organs. Conclusions are rarely drawn by just your eyes, ears, nose, mouth, or hands.

Instead, this data Is sent to the brain to be processed and interpreted. Visual perception of design elements and principles must be understood to create a strong and more effective design.

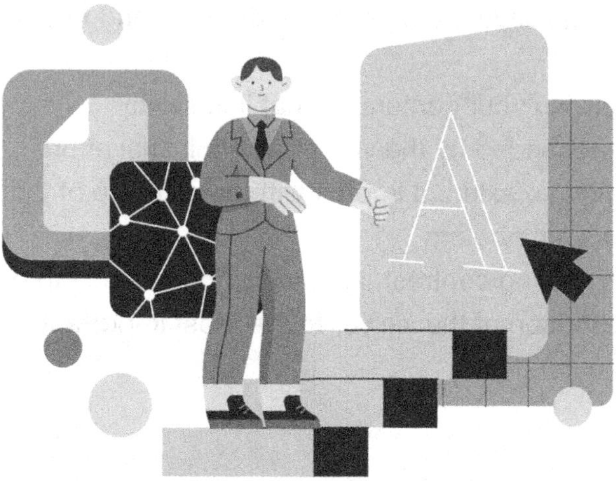

Every design component is a big part of a visual language that transmits messages to an audience. Moreover, understanding how people perceive and process visual information will increase your chances of producing a composition that audiences comprehend and engage with. Let us turn to study our senses and cognitive science to thoroughly understand the science behind visual perception so we can become more sensitive to our designs.

KEY TAKEAWAYS

- The human mind is attracted to visuals and can process visuals faster than processing text. As a result, these visuals have a longer-lasting impact on the brain.
- A visual mode of communication transcends language boundaries and makes processing extensive, boring data accessible.
- In an information-saturated world, a visual helps us convey our information in a fast, efficient way.
- A well-designed visual is a powerful tool that has the potential to change minds and invoke a wide range of feelings.
- Visuals have the power to make our lives easier and better.
- A well-communicating design is clutter-free, organized, and intentional.
- Visuals tend to remove confusion and make data engaging.
- The key parts of the human body involved in visual perception are our physical senses. Our minds process the data collected through our senses.
- Understanding how humans perceive visuals is vital for a designer to create content that speaks through to the audience and is relatable.

CHAPTER 2
VISION

TO UNDERSTAND why visuals are so important, we must first and foremost understand the process of 'seeing' and perceiving those visuals. As we learned in Chapter 1, visual perception is the ability of a human to see, sense, and absorb the information that our surroundings present us with. The process starts when the light from the sun reflects on the objects around us and enables us to see. This is known as vision.

> *"The face is a picture of the mind with the eyes as its interpreter."*
>
> *MARCUS TULLIUS CICERO*

THE ART OF SEEING

What is this vision, and why is it so important? Getting deep into the science of sight, we can see a visual when the reflected light reaches our retina, which consists of around 100 million photoreceptors that absorb the reflected light. The retina then converts this light into electrochemical signals for the brain. There are two types of photoreceptors present in the retina: rods and cones.

Rods are photoreceptor cells found in the retina that respond to low light and are responsible for our night vision. Because they are placed on the ends of the retina, one can learn to improve their night vision by focusing on the side of the object to allow light to reach the rods better. On the other hand, cones are the photoreceptors found in the fovea or central region of the retina. They are designed to handle high-activity tasks, including reading text, perceiving color, and taking in shapes. The fovea plays a vital role in processing visual information by helping us differentiate between shapes, colors, and details. Once this information is accessed by the eyes, electrochemical signals pass via the help of a network of neurons to ganglion cells. This network of neurons has the power to detect any type of contrast and shadow present in the visual recorded by the retina. The fully amended electrochemical signal is then transported to the brain with the help of the optic nerve. This is where the brain kicks in and starts the next stage of visual perception.

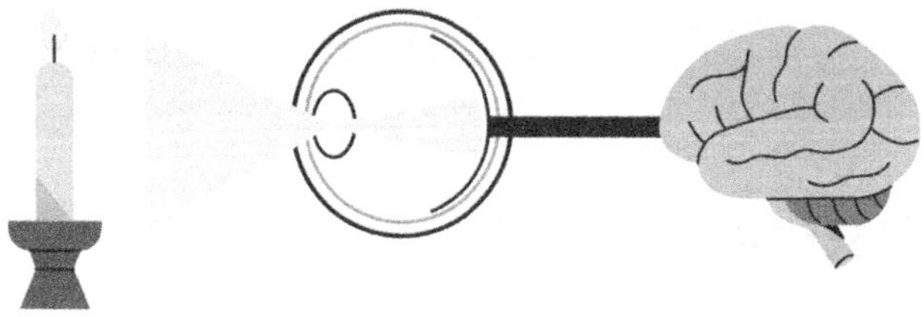

Light passing through the eye to the brain

Despite the copious amounts of visual information we receive daily, our eyes and minds effortlessly process them. Upon reflection, it is baffling how easily and seamlessly this process takes place throughout each day, all our lives. For the most part, you never have to consciously think about how to recognize objects and make sense of things in a 3-dimensional world. Without thinking twice, you instantly recognize objects such as stop signs and traffic lights. This confirms an innate relationship between 'sight' and 'understanding.' The only way to establish and comprehend their relationship is through learning how we see and make sense of the world around us.

The human brain has the capacity and capability to process images actively. Two kinds of visual processing are at play here: Bottom-up and top-down processing. Bottom-up processing takes place at the start, instantly and subconsciously. Initially, when our eyes meet a visual, the brain, in its need to find patterns and meaning in an image, looks for shapes, sizes, colors, contours, and contrast, helping us identify objects and base data important to processing the visual. This data is quickly transmitted to other parts of the brain for analysis.

Now the conscious top-down processing starts. This phase of processing is heavily influenced by our experiences, knowledge, and expectations of the visual. It quickly groups the visual into parts and disregards any information that is useless to us at the moment. Top-down processing plays a huge role in understanding visuals and absorbing the projected data. Although extensive and detailed, all these events occur within a few milliseconds. This is critical to understand because this perception and analysis of visual data explain why it is becoming more and more important to present overly detailed, rigid, and boring data with the help of meaningful, engaging, and interactive graphics.

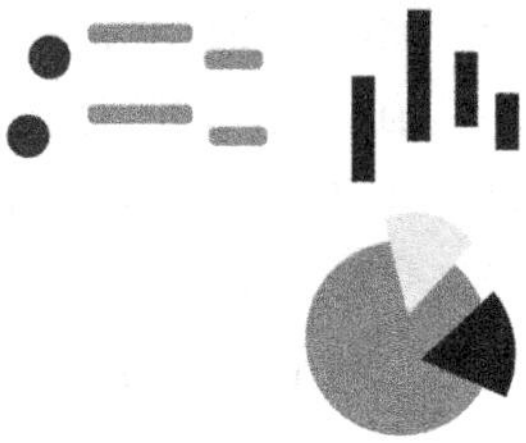

First we identify basic shapes

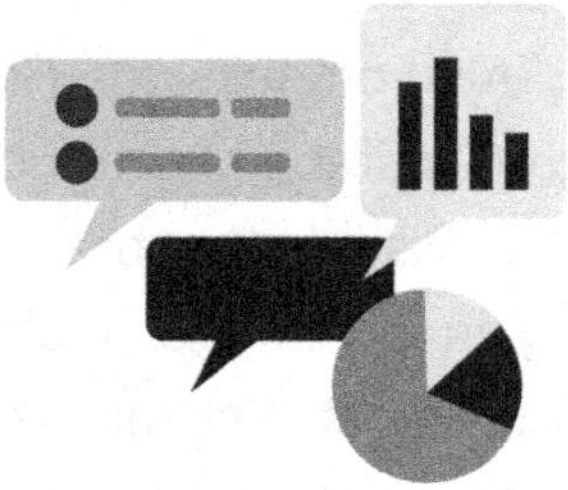

Then shapes are grouped together

These two processes are comparatively independent and can bring different experiences for people. Understanding how to use them to your advantage can help take your visuals to the next level. Why?

Because attaining a general sense of our innate mental and visual abilities can give designers and presenters an edge to express their meanings and messages with an intent that drives through to the user. If you intend to visually convey an important message, understanding how humans comprehend and analyze will help you create a well-designed data visualization that people will understand and relate to. Suppose you intend to invoke a certain feeling, such as curiosity or excitement, in the viewers through your viz. In that case, understanding how emotions work and come into play in the human brain enables you to design consciously and create an extremely powerful visual. Suppose one intends to visualize large amounts of tedious data. In that case, understanding the constraints of memory and how short-term and long-term memories work is an excellent way for you to create a graph, table, or chart that perfectly describes the data story.

Graphs and tables are means of communicating visually. Where graphs are purely visual in nature, tables are partially visual in that we use our eyes to perceive a layout supported by the text. With the help of a table, the text it contains can start to describe a narrative through patterns and groupings. The perfect balance of lines, fill colors, and hierarchy can make them incredibly effective. With a detailed understanding of visual perception, your knowledge of table and graph design will increase beyond the general list of design principles. Armed with this knowledge, you can create sensitive and well-received visuals by the viewers. Only when we apply these meaningful insights into our design process will we be able to present data in a manner that works for people's eyes, makes sense to their brains, and resonates with their feelings.

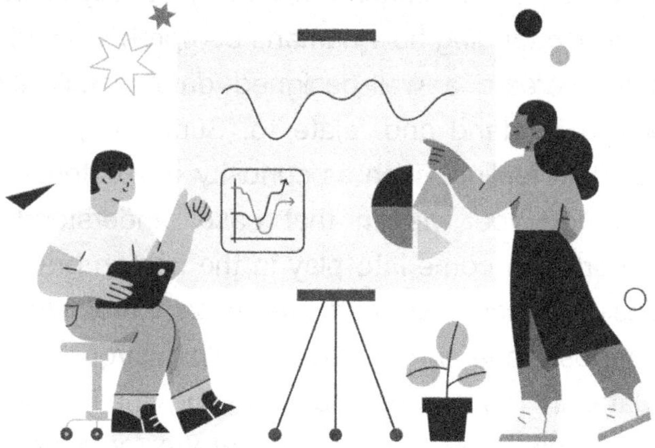

I'm not the only one who thinks so. Colin Ware states in his book Information Visualization: Perception for Design:

> ".....*When data is presented in certain ways, the patterns can be readily perceived. If we can understand how perception works, our knowledge can be translated into rules for displaying information.*"

Thanks to science, the way we see is now well-understood, from the eyes intercepting the visuals, processing them into signals, and then transferring them to the brain to be interpreted in the mysterious gray folds of the visual cortex.

THE EYE VS. THE CAMERA

One excellent way to better understand how a human eye works is to look closely at the process of a camera.

. . .

Our eyes tend to operate a lot like a camera, with multiple components working together to produce an image of an object contained in light. For people who aren't aware of how a camera works, let me break down the process.

The curved surface of the human eye, known as the cornea, is what protects our eyes from the elements. Behind this protective layer lies the iris, a muscle that increases and decreases the size of the pupil, which allows light to enter the eye. When there is a bright amount of light, the pupil becomes smaller and constricts to protect the eye. In low light, it becomes bigger since it needs to bring in more light to see. The iris and pupil work exactly like a camera's aperture, which controls the amount of light let into the camera lens by adjusting its size. Behind the iris is the eye's lens, which tends to thicken when it focuses on objects close by and thin when objects are far away, sort of like the zoom on a camera. Have you ever looked at a distant object and pretended to 'zoom in' by focusing on it? Try it for a fun mental simulation of a camera lens.

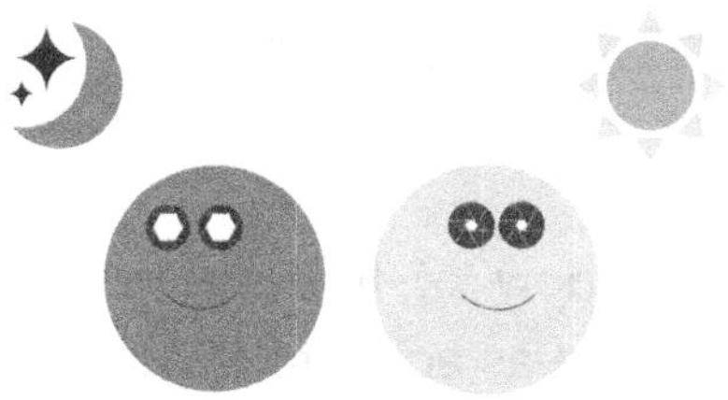

The pupil gets wider in low light and narrower in bright light, just like a camera aperture.

Once light is collected by the eye, it hits the retina (a sensory layer at the back of our eyes) and forms an image. This retina behaves the same way as a camera's sensor or piece of film. It is an essential part of the eye since the majority of information is collected in the retina. Like a camera sensor, an image is affected if this part of the eye is damaged. A basic example is degrading eyesight. It can be said that visual perception is the most developed and efficient of our senses and hence the reason for its importance. Therefore, a camera becomes a perfect tool for capturing our memories, significant moments, and experiences just as the human eyes see and capture them.

Although effortless for us, the process of seeing is not as linear as you might like to think. Our vision is often interrupted by random bursts of information and frequent pauses. This can be explained better by understanding the process of reading. As you read this book, your eyes don't move in a straight line from word to word, left-to-right, or letter-to-letter. Seeing is not linear, nor is it very smooth. According to research, the eye moves unevenly through a text, skipping words while focusing or stopping on other words to better understand what is being read. These jerky and rapid movements are known as saccades. The stops between are called fixations. And when you go back to re-read a word, it is called regression.

During the small window of time while saccadic movement occurs, our eyes are not as acute to what they see compared to when they are still. If anything pops up on a screen or in front of our page during a saccadic movement, our eyes probably won't notice it at all. Despite the many millions of receptors present in our eyes, we can never fully see what is out there, just a portion in specific detail, since some things attract our eyes more than others. If you remember this when designing your visual, you will be capable of creating something that jumps out of the page or screen at the viewer, and the story you are trying to tell will be better understood and perceived.

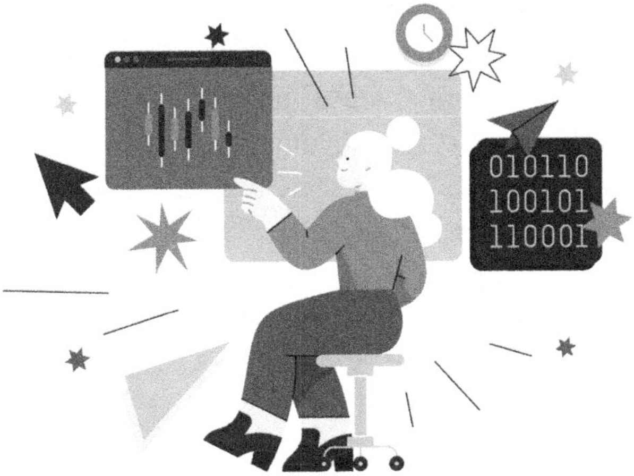

Psychology-driven design has shown great potential in drawing people's attention to exactly where the designer wants. But how do we anticipate where our eyes gravitate to if we don't consciously decide? A careful study of people's habits and past experiences can hint at how to arrange our visuals so the most vital information becomes the part that pops out right away. For example, people customarily read top-down and left to right in the West, while others read right-left in some eastern countries.

Another tip you can try to test whether your layout is effective is to reduce or remove the color from your design and squint your eyes at it. The most prominent elements and groupings will pop out and show where your viewer's eye is most likely to go. Considering psychology while anticipating where people will look will optimize your audience's experience and make the most of your visual.

• • •

Although UX researchers have made significant progress in understanding the science behind 'fast thinking' and visual perception, it is almost impossible to determine precisely where and how everyone's eyes will scan or pause on a page. This is because it varies for each person and is influenced by personal experiences, cultural differences, and the subject of the visual. But doesn't this erratic eye movement makes it easy for people to miss out on details? How can a designer ensure that the user/viewer doesn't miss out on any information? How do you layout a graph or chart in a way that catches the eye and moves it along so that the story of your visual is not only seen but also analyzed and understood?

Well, an excellent way to do that is to create a design structure for all the information so that there is a particular sequence to be followed by the viewer or the human eye. When this pattern or arrangement is decoded, it becomes easy for them to follow it throughout all your other designs. As our eyes move from one visual to the next, consistency in format makes it easy for the viewer to fixate on the most meaningful bits and skip over all the unimportant bits. This way, viewers focus on changes in the data instead of getting distracted by changes in design or composition.

THE IMPORTANCE OF VISION

Learning about visual perception and the human mind is all-important because it helps us learn about the significance of human vision. Using visuals is an accessible form of communication that facilitates people who are multilingual or disabled. It blurs all language lines and makes it easy to understand globally. Have you ever had to resort to hand gestures to communicate with someone who didn't speak the same language? What's the best way to introduce yourself to someone new? A smile. Visual communication allows us to bypass differences in culture and language as it knows no limits of tongue, vocabulary, or grammar and can be perceived by anyone regardless of literacy.

Moreover, precise tables, charts, and graphs can help structure data to make it easy for the audience to absorb and retain facts and figures. It saves them the effort of going through each line of data individually, whether it be a line of text or a column of numbers. Similarly, when we think of initiating a call to action, we know that a powerful, meaningful visual will reach a greater audience and invoke a broad set of emotions within them... something that text alone will have a hard time accomplishing. A graphic can be more comprehensive and easier to understand. It can grab a user's attention over text, even if it is poorly designed. The power of visual communication is infinite and can be easily controlled by designers for their purpose.

Understanding human vision will help you create powerful visuals. It will help you connect nicely with your audience because you will know what they relate to and be able to reach their emotions. You will be able to draw connections between how people see a visual and the principles of design. Combining them will create strong layouts with a more powerful message behind them.

In so many words, understanding vision is a significant part of leveling up your design that will encourage, motivate, educate, and help people resonate with your visual creations. You will know how to direct someone's eye gaze to exactly where you want and connect them to your data story.

How does vision feed into our memories? How do humans retain this memory, and what makes a memory meaningful? There is a science behind this too. In the sensory and memory store, information arrives from five different senses, but this information is temporary and is lost quickly. Sensory information is fed into the brain, where it encounters three types of memory: Sensory, Short-term, or Long-term. Attention becomes the crucial step in remembering something. When a person focuses their attention on a scene, what they see is sent to the short-term memory where their brain interprets it. If they are not really paying attention at the time, the scene slips from short-term memory and is forgotten instantly. This is why grabbing attention is important! Our brains are problem solvers and want to understand the world, but they are also quick to conserve energy.

When you face an incredibly powerful visual that invokes a strong emotion, you are bound to remember it for a longer time. This is where the expertise of a designer comes in. Suppose you create a powerful, thought-provoking, and informative visual during the design phase. The brain is bound to remember it and move it into working memory and potentially long-term memory. This is the next step of the visual perception path. From here, we start making sense of everything we "see."

KEY TAKEAWAYS

- When light reflects off objects and hits the eye, that is when an image is formed on our retina. Seeing an image form on the retina is known as vision.
- The retina is the most sensitive part of our eye and has over 100 million photoreceptors that enable us to see.
- Our eye behaves precisely like a camera and enables us to record our most cherished experiences in the form of memory, just like exposing a photo on a piece of film.
- The process of seeing is not as smooth as we would like to think. Instead, the eye makes erratic movements while taking in visual information, known as saccades.
- Saccadic movement can often skip over details. But this can be rectified through good design that makes the eyes move towards the important part of the visual for maximum impact.
- Using visual communication is an excellent way to transfer information across the boundaries of language and culture. It helps people connect regardless of their literacy level.
- Understanding vision helps you understand how to direct the viewer's eye and have more control over how your visual is read.
- What is collected by the eyes is transported by neurons and processed by the brain. After a visual is processed, it either becomes a memory and stays with you forever or decays and is forgotten in a few minutes.

CHAPTER 3
MEMORY

WHILE WE ARE TALKING about visuals and visual perception, we must not forget to study the science behind the act of visualizing. Some might call it imagination, daydreaming, or even manifestation. It all comes down to what your eyes see, the brain perceives, and the mind understands at the end of the day. So, what is visualization? It is a term that has been popping up around us a lot. We hear the phrase in wellness practices, social media, and even athletics. One meaning found in the dictionary is "the formation of a mental image of something".

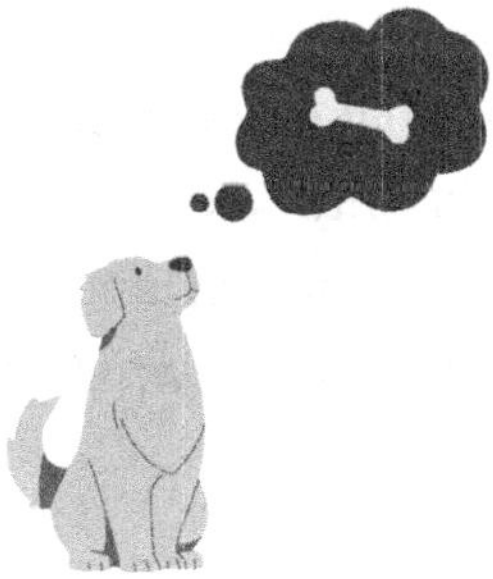

Manifesting

Unfortunately, that does not explain how it works and integrates with design. Visualization, in simple terms, means picturing in your mind's eye what you want. This can be anything. In the world of art, visualizing means imagining the final result of your artwork/design before implementing it in real life. Someone might stare at a blank page in their sketchbook and think about materials to use or themes they might explore. If you are striving to create a powerful, data-driven visual, the first step in the design phase would be to imagine what the visual space will contain. Are there specific shapes, chart types, or colors that come to mind when you think of your data story? Are you trying to keep it simple, straightforward, or combine layers of complexity? The visualization process is often involuntary. We do it constantly without ever 'thinking' about it.

THE SCIENCE BEHIND VISUALIZATION

Understanding the science behind visualization is necessary to figure out why only certain things grab our attention and stay in our memory despite the multitude of things that pass in front of our eyes. And even then, only a certain amount of information gets retained in our minds.

Our eyes don't do most of the seeing. It is actually our minds that 'see.' Our eyes are only responsible for recording reflected light and transmitting it into electrochemical signals. These signals are sent to be understood and perceived by the brain. Whatever goes on in our brains is what we experience as visual perception or 'visualization.'

. . .

The science behind visualization is related to the neural patterns inside our brains and how they respond to the process. The nervous system of humans is stimulated just by the mere thought of something, even if they haven't experienced it before. When you sit down and try to visualize an image, a memory, or a desire, you create neural patterns in your brain. These neural patterns respond as if you had actually stood before the scene, experienced the event, or achieved your desire. Athletes or entertainers often run through their future performance in their heads as a way to mentally train alongside their physical training. This potent phenomenon can help you during the design process as you try to understand how your audience's minds will react when they come across your visual. It is also proven to boost productivity as having a clear direction helps you feel calm and focused. Practicing visualization is an excellent tool to utilize and can help you make a big step toward delivering better results to your audience.

Consequently, the brain is the superpower that does all the 'seeing' and perceiving of the external world's constant stimuli.

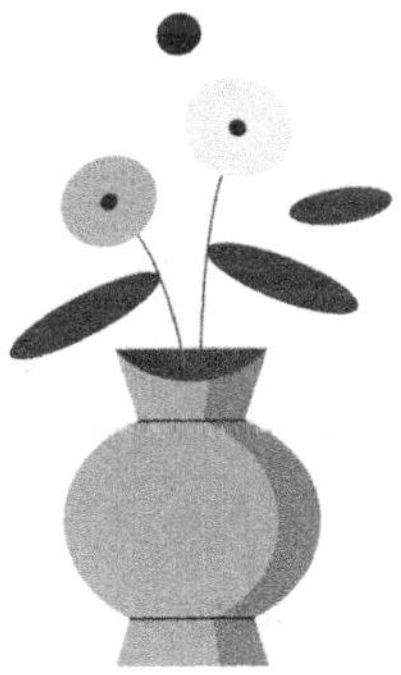

Let's look at it this way: when you look at a bouquet of flowers, you are immediately pleased with the color, shape, and grouping of the blooms. This is because the light reflecting off the flowers reaches your eye and generates a signal that triggers a special kind of memory called buffer memory. This is a temporary location. Here the signal is processed and sent forward according to its impact on the brain. The next stop is the working memory, where the signal is analyzed, transcribed, and evaluated. At this stage, the brain tries to relate what it is processing with previous knowledge. Finally, if the information is repeated or considered important for future reference, it is moved into permanent memory.

This phenomenon is widely known as the multi-store model of memory proposed by Richard Atkinson and Richard Shiffrin. This structural model explains that three types of stores make up memory: a sensory register, Short-term memory (STM), and Long-term memory (LTM).

SENSORY MEMORY: THE MAIN GATE

Each day, we absorb such an extensive library of information through our eyes, ears, nose, and mouth. This is known as 'Sensory Memory.' All information collected through sensory information cannot be stored or registered permanently. When we focus our eyes on something specific, data is transferred from sensory memory to short-term memory.

Most of us rely on iconic memory to receive and respond to the stimuli around us, also known as visual sensory memory. Iconic memory stores visual data after the trigger has been removed or ceased. Although this memory has a considerable capacity to receive stimuli from the environment, it declines rapidly. Any information collected through our eyes only stays for half a second, depending on the object's brightness. For example, if you are sitting in your living room in the warm evening light, the things you see will disappear faster than an object you see on a bright, sunny day. This is also why after you look away from a high-watt bulb, its shape still hangs in your vision.

One thing to note is that this type of sensory memory is automatic and subconscious. But what does this mean? It simply means that details gathered through our eyes are processed within milliseconds before we know it. These kinds of attributes are also termed pre-attentive attributes. The processing of pre-attentive attributes happens at a very high speed and tends to detect only a specific set of characteristics. This type of data processing is crucial for you to understand when designing, as most of the visuals you create tend to go through pre-attentive processing. An easy way to comprehend why our brain goes through this is by studying the brain model provided by Daniel Kahneman. According to Mr. Kahneman, the brain is divided into two systems:

System 1 - fast, automatic, unconscious, and stereotypical.
System 2 - lazy, slow, conscious, and analytical

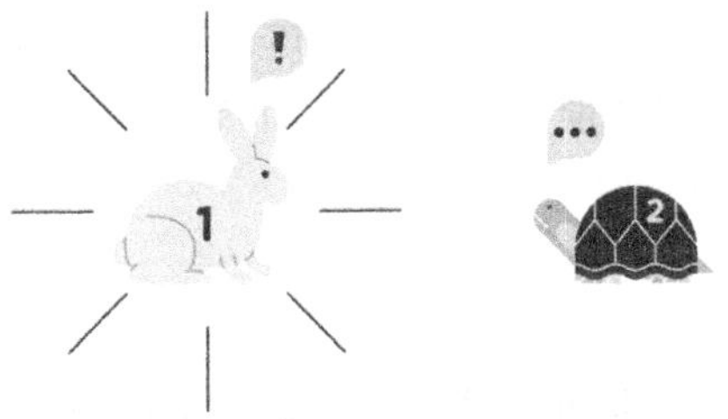

System 1 vs System 2

Since we know that a human being collects and processes a high amount of data within milliseconds, it becomes impossible for the brain to analyze everything consciously, attentively, and slowly. Attentive processing would take too much time, drain too much energy, and render us unable to multitask. Therefore, both systems have evolved extensively through time and are necessary for our survival.

System 1 is connected with pre-attentive processing and hence is the most significant system for you to familiarize yourself with. You can define pre-attentive attributes as the elements your eyes are attracted to when you first look at a visual. This is important for you to understand because when you master pre-attentive processing, you can control where your audience's eyes initially go when they look at your visual. In a way, it enables you to control the visual narrative.

With the help of these pre-attentive attributes, you can learn to highlight the most critical parts of your visualization. The narrative you control through pre-attentive attributes determines what story you want to tell, how you want to tell it, and the impact it generates on the viewer's mind. This is the most crucial step in your journey to make excellent visuals. This process lets you bring your audience to instantly grasp your visual's road map and intention.

If you are ready to learn all the ways your design can have a meaningful impact through pre-attentive attributes, let us look at its three main categories:

COLOR

Color has to be one of the chief elements in an impactful design. It is one of the simplest ways to attract the viewer's eye to something significant. Unfortunately, it is also easy to go too far with color. There are two ways you can use color for an impactful visual: hue and intensity.

Hue can be called the general use of color in contrast with a monotone (black and white) base. Color can be used in graphs and charts, illustrations, and other types of design to draw the viewer's attention to a specific element. In a neutral home interior, colorful art or decor can create focal points in the space. In a chart or graph, the data you want your audience to see first can be in color, while the rest is black or grey.

Using color to bring attention to the focal point of a visual enhances your audience's chances to pick up your meaning quickly. The trick here is to use the 'right amount of color.' For example, you can choose different colors for positive or negative changes in the data, like with a heatmap. Emotionally warmer colors such as red and orange can indicate rising values, whereas cooler colors like blues or greens can mean decreasing values. Using different hues where you want to note important details can make all the difference. The user will have a strong grasp of the concept without even trying. And a design that reaches people more efficiently in less time is what we aim for.

. . .

Controlling the intensity of the color is also an excellent way of determining that your audience receives the maximum information with minimal intervention. The intensity of a color refers to the brightness or darkness of the hue. You can think of saturation versus desaturation. This can be done by controlling the opacity of the color and using it to highlight important or most recent bits of information in a chart. Referring back to a heat map, you can also communicate differences in temperature using a single color but varying the darkness and lightness corresponding to values getting higher or lower.

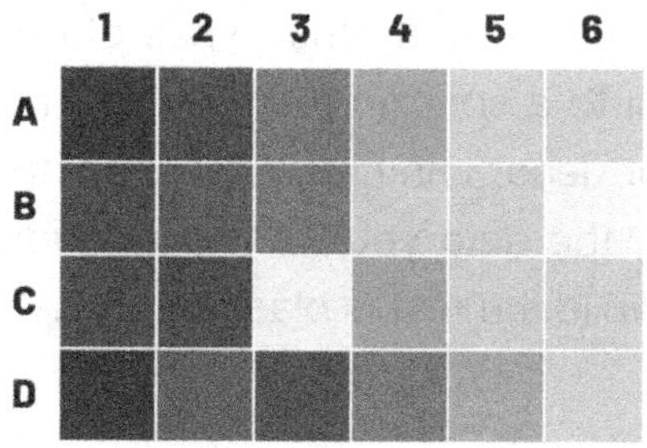

A heatmap uses varying color intensities to show changing values.

FORM

Another pre-attentive category out of the three is Form. This category includes the most amount of the pre-attentive attributes that can be used for an impactful design. These are curvature, length, shape, enclosure, width, and size.

Let's talk about these attributes in order.

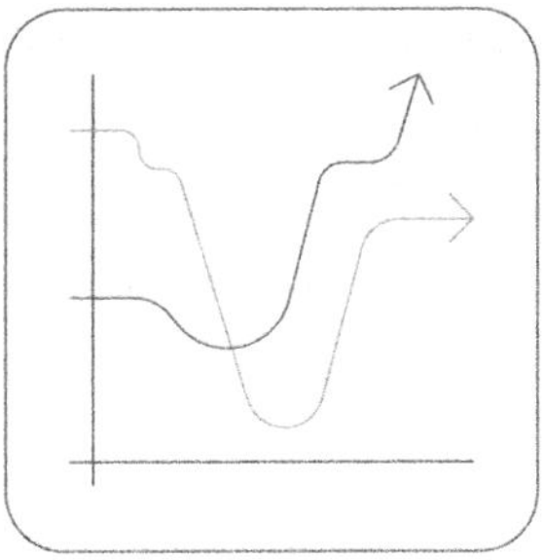

Our eyes pick up curved lines faster than straight ones.

Curvature is essential to use in a graph to depict how rapidly the changes are occurring, and it is more readable by the human eye than a straight line. It is an asset to graph making and is automatically implemented when generating line graphs. Curvature is incredible for visualization.

The second attribute, **length**, is also a noteworthy thing to consider during design, especially if you are prone to working with graphs and a huge data set. The human eye is quite susceptible to detecting spatial distances. It has the ability to distinguish between different lengths, which explains why the bar chart is considered a valuable and go-to tool to compare quantities of values. One tip to further enhance the effect of this attribute is by adding color.

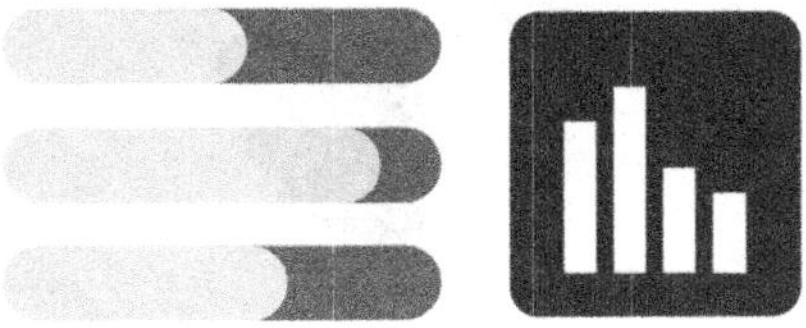

Bar charts utilise Length to compare values.

Shape is another attribute that can significantly enhance a visual. Whether used to mark positions in maps, create a focal point in an illustration, or distinguish between certain segments or groups, a shape can play a huge role in directing the user's gaze towards the desired object.

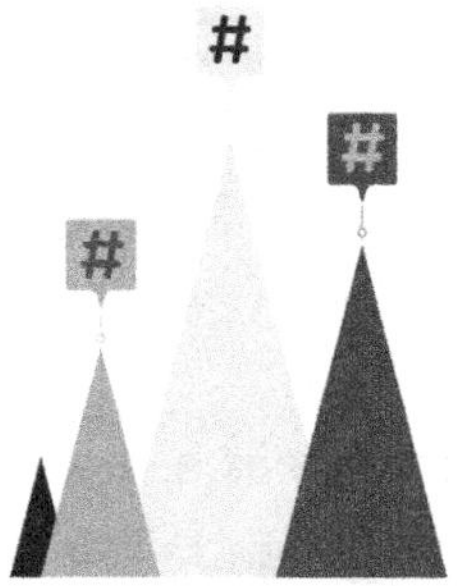

The squares draw the eye to the values at the peaks of the pyramids.

Enclosure means that any object surrounded by color will be more visible and eye-catching than an object that isn't. Think of an outline around a shape or text. This technique ensures that the visual doesn't become too cluttered. Use it sparingly and on what you want to pop out most. This is because if you enclose every object in one graphic, it would be highly overpowering for the audience to look at.

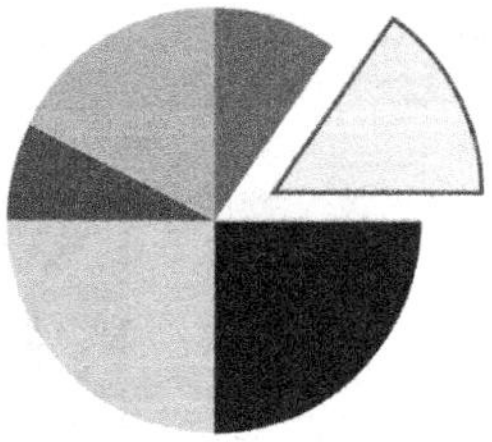

The outline around the pie section catches your eye right away.

Size, next to color, is the most distinguishing attribute commonly used to make a visual pop. The difference in the size of objects can have a drastic impact on the eye of the viewer and can direct their eyes in a sequence set by you as a designer. It can also communicate value without needing text. For example, we may look at two circles of different sizes and assume the larger one represents a greater quantity of something than the smaller one.

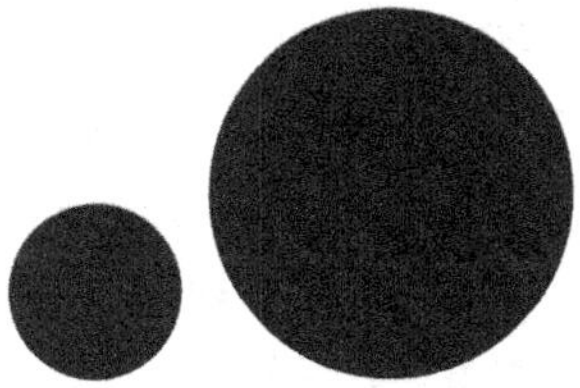

The bigger circle makes you think greater value.

POSITION

The third category that we need to acknowledge is position. The human eye has an excellent ability to detect shapes, patterns, and arrangements. This can be exploited by those who want to direct the user's eye to a specific area before letting it roam around on the visual. So whether you design graphs and charts or posters and illustrations, learning how to direct the user's gaze to a specific important element in the layout can be beneficial in controlling the narrative of your story.

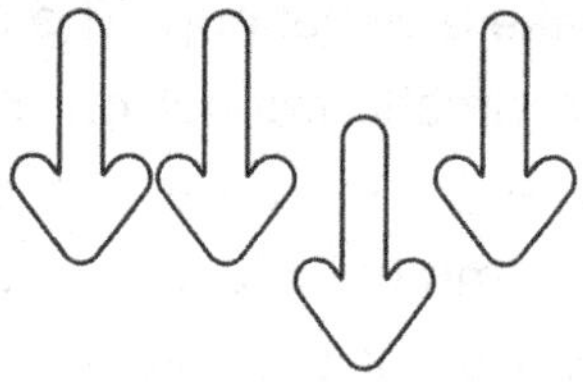

Which arrow did you notice first?

We are constantly surrounded by these attributes that contribute to a successful design. Yet, a generic viewer cannot tell you why a particular visual appeals to them more than the other. This is because the human brain is constantly doing our processing and thinking for us. When we look at a graphic or visual, our eyes and brain automatically coordinate to direct our gaze first to something placed in a contrasting color, shape, size, or position.

Learning these attributes and the science behind them tends to influence your design decisions because they teach you how to design the kind of visuals that grab your audience regardless of whether they understand the science in the background. So paying attention to minor details like color, contrast, shape, and object location might seem insignificant, but they are actually quite important.

WORKING MEMORY: THE PROCESSING CENTER

When you talk about 'Working Memory,' it refers to the process after which sensory memory takes the data collected through our sensory organs (visual, auditory, gustatory, and olfactory) and converts it into electrochemical signals. These signals are transmitted to the working memory, where they are processed and evaluated based on our current understanding and previously stored memory. All the gathered data from our sensory organs are combined into visual chunks. These big information chunks are processed and analyzed by the conscious mind in the processing center. At a time, the human mind can handle around 5 chunks of information before it loses attention or needs to reprocess the data. Pieces of information designed to form a large, coherent pattern can make it easier for the user to remember and hold in working memory.

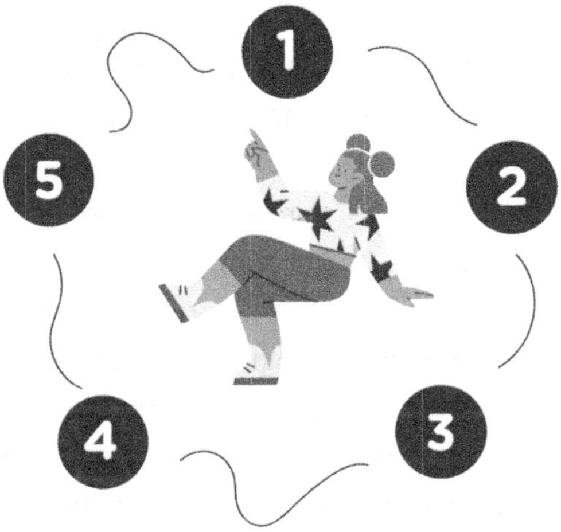

The working memory becomes a place where perception takes place. The data collected from our senses is pieced together and compared to existing information stored in long-term memory. If the information matches any preexisting knowledge, the brain finds it easy to understand and make a connection; if not, the brain makes guesses as to what it sees. Have you ever traveled to a new city or country and compared the buildings or food to your home town? Our brain always starts by comparing new data with what we already know to establish a baseline of understanding.

The information gathered in the working memory is temporary and has low storage capacity. The period for a piece of information to remain in the working memory depends upon the number of times it is rehearsed or revised. How many times did you have to sing the alphabet song as a kid? Too many. Why is it that you can bust out the chorus when that one song you haven't heard in forever pops up? Because these songs were repeated over and over in your mind.

Memory only lasts a few seconds if it isn't repeated and then vanishes right away. Since the information in the working memory stays for only a short period, it is also known as Short-term memory or STM. The duration of short-term memory is between 15 to 30 seconds but can be stored for longer with constant revision. However, the information collected in the processing center does not remain there. If any new information needs to be processed, the old data in this memory type must be moved to long-term memory or forgotten completely. This information is fundamental for you because the way a visual is designed can affect whether your viz is memorable or not.

Arthur Conan Doyle's famous literary character Sherlock Holmes purposefully avoided keeping any information that he deemed unimportant in his memory. He said anything that is not useful for solving crime is a waste of space and mustn't be given any room in his mind. Doyle's character believed that the mind is a vessel, the walls inelastic and cannot be distended.

Therefore, keeping all the information you encounter does nothing but clutter your mind and devoid it of space for crucial knowledge. This is why the ingenious Sherlock Holmes had absolutely no clue that the Earth revolves around the sun:

> *"It is of the highest importance, therefore, not to have useless facts elbowing out the useful ones."*
>
> *"But the Solar System!" I (Watson) protested.*
>
> *"What the deuce is it to me?" he (Sherlock) interrupted impatiently; "You say that we go round the sun. If we went round the moon, it would not make a pennyworth of difference to me or to my work."*
>
> —ARTHUR CONAN DOYLE, SHERLOCK HOLMES: THE ULTIMATE COLLECTION

Working memory or short-term memory is not only responsible for holding the data; it also organizes the data, determines the foreground of the visual from its background, groups different elements together, and collects textures into basic forms. It does all that needs to be done to understand and fully comprehend a visual.

LONG-TERM MEMORY: THE ARCHIVES

Long-term memory is the final storage place of information, whether for a few minutes or a lifetime. This memory type is a bit mysterious in that we don't fully know its limits. We have the potential for an endless capacity for storing memory. Any memory that reaches the long-term memory (LTM) has the potential to be recalled or pulled up at convenience and stays with you for an unlimited time.

Long-term memory is divided into two types: Explicit (knowing that) and implicit (knowing how). The difference between the two types of long-term memory is that one is procedural knowledge, and the other is declarative knowledge. Procedural knowledge means knowing how to do things. For example, if you know how to play guitar or any other musical instrument, it is included in the implicit long-term memory and does not involve any conscious thought on your part. This kind of LTM is unconscious and automatic. On the other hand, declarative knowledge means knowing what things are. For example, all the factual information like knowing London is the capital city of England or that bears are mammals is included in explicit LTM.

Memory is the critical tool for determining and analyzing our interactions with our surroundings. Without memory, we cannot and will not be able to make sense of anything around us, and it will render us incapable of interacting with our environment. But memory is not as straightforward as we expect it to be. Memory is hugely complex and is constantly surprising scientists. It is continuously evolving and is greatly impacted by our surroundings. Let us attempt to understand what memory is and how it affects design.

The total capacity of what a human brain can and cannot store hasn't been identified completely. But the data received, perceived, and analyzed is stored as a pattern of neurons. The stronger the imprint of the neuron, the longer the information stays in our memory. These imprints can be made stronger by repeating and revising the information over some time. While designing, we must keep this in mind whether we create a visual intended to be stored in the LTM or temporarily in the STM.

For example, are you presenting a quarterly update to stakeholders? This visual would most likely be based on a smaller data set and provide quick information about a current state. You would want to focus on keeping the audience's attention with pre-attentive attributes and providing a straight-forward story that their working memory understands. Or are you creating a visual for a non-profit to attract more public funding? In this case, you'll want to connect emotionally and be memorable. Using repetition and focusing on your message will stick in your audience's memory.

Understanding your intention will help you make design decisions. This includes choosing the best chart or graph to use and how to layout your visual elements in a way that engages the STM with a better chance of being remembered long-term. However, designing so that your audience moves your visual to their LTM requires a great deal of understanding and can be extremely difficult. Try experimenting with the arrangement of your graph components using pre-attentive attributes. You can create different versions of your visual and get feedback from your peers on which is the most successful.

. . .

Humans have not only limited space in their STM for information but also limited attention spans. The attention spans drop if a visual is cluttered, too noisy, or impossible to read. The aim, therefore, should be to keep the process of data perception and memory in mind before you design so that you can hold the attention of your audience for a greater period. Moreover, the audience is more likely to retain information from a visual if it is made easy, accessible, and organized. Remember that we process information in chunks and can handle about five at a time. These chunks can be designed effectively to imprint data in the user's mind enabling them to better understand the visual in less time. They allow for better and more effective processing and can be done by arranging distinct groups of similar content together. For example:

- Chunking text of similar context into paragraphs
- Grouping visually similar items together
- Grouping together strings of alphabets or letters

This book gathers multiple guidelines and design principles that will help you convey your message to the audience while having control of the narrative. These are, of course, not strict rules or guidelines that must be followed, but with the understanding of memory in mind, you are more equipped to design an impactful visual for the audience. With this understanding of how the mind and memory work, we can design a visual that is not only fun to look at but also easily digestible.

This also means that by understanding how your audience sees and perceives things, you put yourself in a better position to communicate information effectively. A few tried and tested methods that can be incorporated to make design better are:

- Grabbing the audience's attention by using the three pre-attentive attribute categories in your visual: color, form, position
- Optimizing comprehension times by using strategies of contrast, shape, and line to guide the viewer's eye to the most meaningful bits of information
- Defining patterns to help trigger the working memory ultimately makes the visual more pleasing, interactive, and fun for the viewer
- Connecting with the user on an emotional level

All of these methods are great ways to ensure that your design is appealing and memorable. People tend to gravitate towards designs that relate to them emotionally. A positive emotional experience is more likely to promote healthier, more efficient memory storage. In contrast, a poor experience has few to no chances of being remembered in a way you want. The human mind tends to respond to emotion well. So, if we design a visual in such a way that the experience of the audience is positive and happy, they are more likely to remember your work in the long run.

KEY TAKEAWAYS

- Visualization means training your mind to picture whatever it is that you want.
- Eyes don't do the actual seeing; it is our minds that 'see.'
- When you visualize something, the neurons in your brain create a pattern. This neural pattern behaves the same way as if you have experienced that event.
- There are three types of memory: Sensory, Short-term memory, and Long-term memory.
- Sensory memory includes all the data collected through our five senses.
- There are two types of processing systems in the human mind: System 1 is fast, automatic, unconscious, and impulsive, and System 2 is slow, calculated, conscious, and analytical.
- The sensory data is taken to the processing unit or the Short-term memory.
- Information stays in the short-term memory for only 25 to 30 seconds unless repeated.
- Information that is termed useful to keep travels from short-term memory to long-term memory and the rest is forgotten.

CHAPTER 4
EMOTION

UP TILL NOW, we have been viewing design in terms of visuals that have a particular physical impact on its audience. We started the process of understanding a visual by learning about the science behind how we see, think, and remember. These things are studied in this order because that is how an audience perceives a visual. To review, light hits an object and enables our eyes to record the signal and send it to the brain. From there, the signal stays in the mind until it is evaluated by the brain. If the working memory creates a significant association with the visual, it tends to stick as memory or thought.

So, what part does emotion play here? According to a study by Daniel Kahneman, humans are not purely logical beings and are driven to certain tasks by our emotions. This is because our feelings significantly influence our decisions, attention, and memory. A design that is based on emotion or triggers a sensation in the audience can greatly impact the viewer's overall experience. There are two general approaches that you can take to design an emotionally engaging visual: the first would be to target the aesthetic appearance, and the second would be to include interesting and impactful interactions with the viewers.

One thing that experts in the field of design stress most is how well your design relates to your audience. If your designed visual invokes a positive feeling, you will have a greater chance of speaking through to them. It is pertinent to remember that people are more likely to forgive your shortcomings, like your design, and relate to it if it rewards them with positive or personal emotion. But how can a design be based on emotions, and how can understanding emotions help you inform your design choices?

DESIGNING FOR EMOTION

As with all the information we have gathered so far, paying attention to the science behind human emotions and reactions helps design a visual that communicates strongly and passionately with an audience. It enables you to cater to a specific audience and create a long-lasting impact using their emotions. Of course, it's not quite possible to design for everyone as each person is unique and responds to visuals differently. However, suppose you focus on what you need to portray with maximum impact and minimal words and information. In that case, you will become an exceptional designer. Ultimately, who gets to decide if the design is good? Your user, your audience, and the people that receive and understand your message through your design decide.

Using minimal information to portray your idea and intention is crucial because focusing on too many details will make the design hard to organize and look cluttered. You will risk losing the attention of the user or audience. You must remember that your layout is not supposed to be a puzzle. Instead, it should be the solution. In other words, your viz shouldn't create problems but help them solve them.

Emotion is not a complex concept to grasp. All of us feel some kind of emotion and tend to connect or resonate with other people based on these feelings. A visual that lacks emotional connection not only lacks the 'aha' element and the impact but also lacks a clear story, which ultimately makes it an unsuccessful design. The first step to designing a good visual is to narrow down the most relevant parts of your data you want to communicate. Once you have decided what your narrative will be, map out how this story will unfold. The next and most essential step is to think about the viewer's emotional journey. This means thinking consciously about how people will feel before viewing your visual, how they will feel in the middle of studying it, and how you expect them to feel after the interaction.

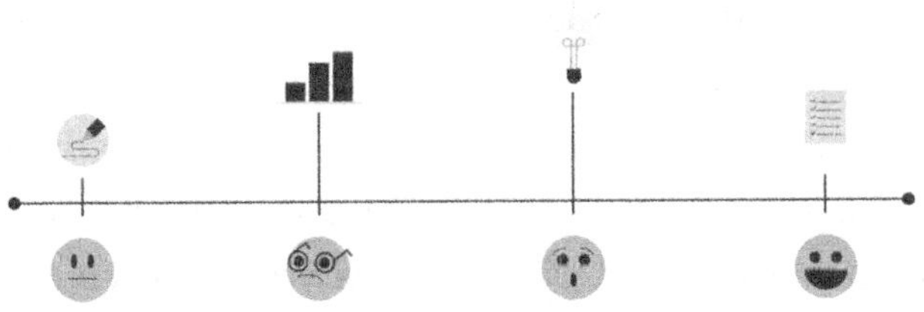

A successful story can inform, educate, inspire, and drive your audience to take action

If your audience has a positive reaction to your design, it will leave its mark on the user, perhaps even transporting the visual to the long-term processing unit in their brains. This positive impact will also trigger your audience to focus on and want to learn more about what you are trying to communicate. Whereas, if the visual negatively affects the audience, such as making them bored, confused, or even offended, they will take a step back and disengage with your design. And if they are forced to interact with it, the negative experience of the first design will stick with them, and they will develop a bias toward your work. This means that not only could you lose your chance to convey your ideas and thoughts at that moment, but your boss or client might not want to give you another creative opportunity in the future.

But this is not all that is required for a good design. If it were, the book would conclude here with a happy ever after, but there is much more to learn. Using and studying human emotions to design is undoubtedly a great idea, but it also has its set of risks. Suppose you focus too much on emotion and not a lot on functionality, usability, or consistency in the visual. In that case, the positive experience you worked so hard for will vanish and turn into frustration. The key to a functional, meaningful, and influential design is to work out your intention for your audience. What do *you* want them to feel, learn and understand from your work? Try to set up a list of incentives you want your audience to achieve, such as:

- to be informed,
- discover something new
- take action
- be entertained

If you work through these incentives, it will be much easier to design a visual that speaks to you and your user.

BREATHE IN, BREATHE OUT

Data authors often skip out on pairing the incentive with the emotional aspect during the design phase of a visual. Instead, they tend to focus on one attribute or the other. But design is hardly ever this simple. Instead, design is often thought of as indulgent frosting, an afterthought. But unforgettable designs are those that consider the aesthetics along with the functional aspect of the design.

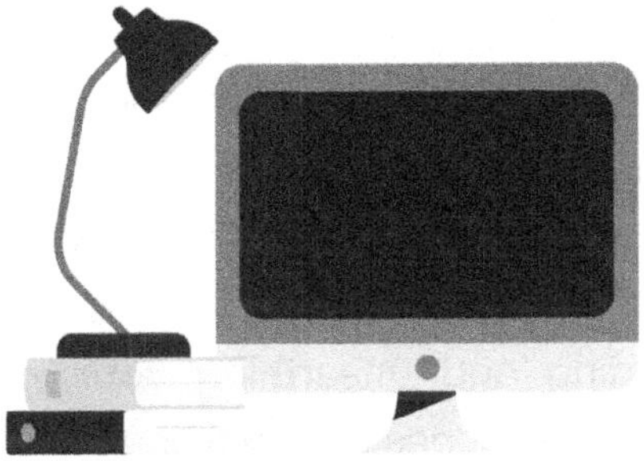

Let us look at how Apple planned their "Breathing Status LED Indicator" for the Mac in 2002. The status light on the front of their desktops and laptops was designed to gently pulsate while in sleep mode. The Apple designers went in-depth to study the science and considered the context within which this light would be seen or experienced by the user. They found users would be mainly in a dark office, bedroom, living room, or workspace where the light quality would be dim and soothing, or the laptop's light would be the only light source.

. . .

The way the status indicator pulsed was especially specific. It was designed to mimic the breathing of a human body at rest, around twelve to twenty breaths per minute. This gentle pulsing created a rhythmic, soothing environment that relaxed the user. It produced a subtle, calming effect on the user without being an in-the-face experience. Users often didn't notice it happening because the design was so calculated and in tune with the requirements of the human brain. This can correlate to the science behind the Golden Ratio present in nature and what we see mimicked in the Parthenon's facade in Greece. The human eye may not comprehend why the pristine white columns are so pleasing to the eye, but the design will have achieved its goal of making the structure look aesthetic.

This kind of sensitivity in design becomes crucial to creating next-level, meaningful, and impactful experiences for the audience. An intelligent design like Apple's Status Light encourages you to see yourself in your work and imagine the user experience. Understanding and unearthing your audience's deep-seated goals, fears, and desires can enable you to design an extraordinary and impactful visual. Only when you put yourself in your audience's shoes can you create visuals that encourage and inspire people to engage with your data, feel excited about what they see, and remember it long after.

KEY TAKEAWAYS

- Humans aren't purely logical beings and use emotions to help them make decisions and navigate life.
- Our emotions have a significant influence on our decisions, attention, and memory, and therefore understanding human emotions aids in designing better visuals that speak to the audience.
- Learning the science behind emotions can help you design a visual that speaks more with less.
- A visual that incites a positive reaction in the audience encourages curiosity and a deeper look at the design. Still, a graphic that creates a negative experience forces people to turn away from the design and mistrust the designer.
- Having a solid mapping of your story, the narrative you want to choose, and the emotional reaction you want to instigate in your user is an excellent approach to a good design.
- Apple's "Breathing Status LED Indicator" proves how good design can be subtle and fulfill its purpose with minimal effort by the user.

PART TWO
ELEMENTS AND PRINCIPLES OF DESIGN

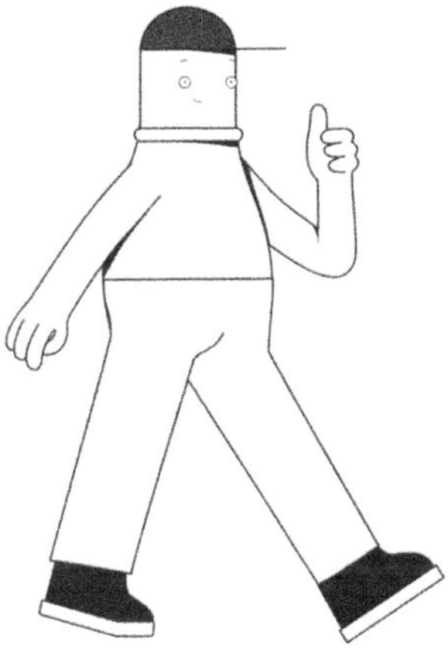

"Good design is obvious. Great design is transparent."

JOE SPARANO

CHAPTER 5
DESIGN TO INFORM AND COMMUNICATE

IN THE LAST FEW CHAPTERS, we learned about how a person experiences a visual and the fundamental science around it. Now you understand how a visual impacts your audience. Learning about the mechanics of perception is highly crucial; still, it is high time we learn about design, its types, and how to go about it. There are multiple fields of design, from interior to graphic design, as it is a broad domain. But, in this chapter, we will pay attention to two familiar types: infographics and data visualization.

Have you ever found yourself incredibly lost in a new space, such as an airport or a foreign country? Have you looked at an instruction manual for a device wishing it was easier to understand? We all have had one or more experiences where we found ourselves looking for a signal, a visual aid, or a symbol that could guide us. Unfortunately, these situations often lead to anxiety, frustration, and sometimes fear for most people. It is in those moments that visual design can come to the rescue.

WHAT IS INFORMATION DESIGN?

Visual communication design, often known as information design, is a way for an individual, object, etc., to present information in a clear and accessible manner. In simple terms, it means the successful curation of a document, including the text, layout, font, format, color, etc. It helps users understand complex information by breaking it down into a simple visual format, which is convenient and time-saving while transcending language barriers. It is challenging to associate only one definition with information design because this design field is so diverse.

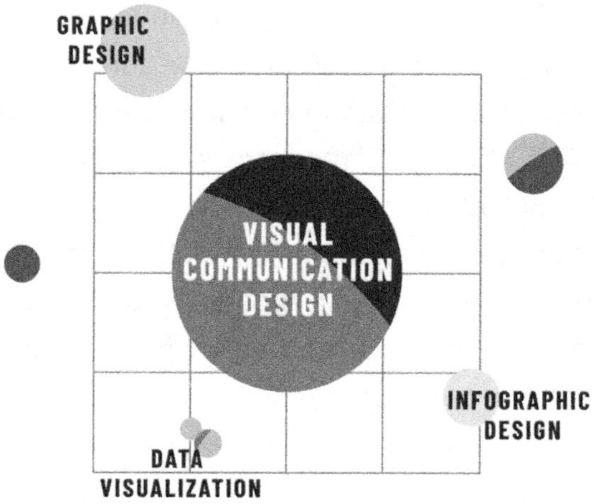

Over the years, the field of information design has evolved into various overlapping branches. Someone who doesn't have preexisting knowledge may get confused when learning that graphic design, information design, and data visualization are progressively more specific types of design under the umbrella of visual communication. Graphic design represents design visually in a 2D space. It communicates using the elements and principles of design, such as layout and typography. Generally, graphic design practices creating strong visuals as a way of storytelling or problem-solving. On the other hand, information design gets more specific and creates visuals to share complex information by combining data, text, and illustration. This kind of design is used in marketing, advertising, or website design to potentially inform, persuade, or entertain.

THE IMPORTANCE OF INFORMATION DESIGN

Information design is found everywhere. Take a look at street signs, websites, magazines, menus, templates, etc. It is impactful for the following reasons:

- It helps condense lengthy text or information in a simple, understandable manner.
- It gives people direction and helps them navigate the potentially confusing world of directions, facts, figures, and commands.
- It conserves time and energy.
- It helps users be efficient in finishing tasks and solving problems.
- It ensures that content is delivered in a way that resonates with people.
- It uses visuals that enable a larger audience to understand and access valuable information.
- It accommodates non-native speakers, people with low literacy rates, the mentally challenged, and the visually impaired.

Information design enhances our comprehension by creating strong visuals that emphasize critical content. There are generally two approaches to information design: creating infographics or data visualizations. Even though these are two similar types of visuals, they are slightly different.

WHAT ARE INFOGRAPHICS, AND HOW DO WE DESCRIBE THEM TODAY?

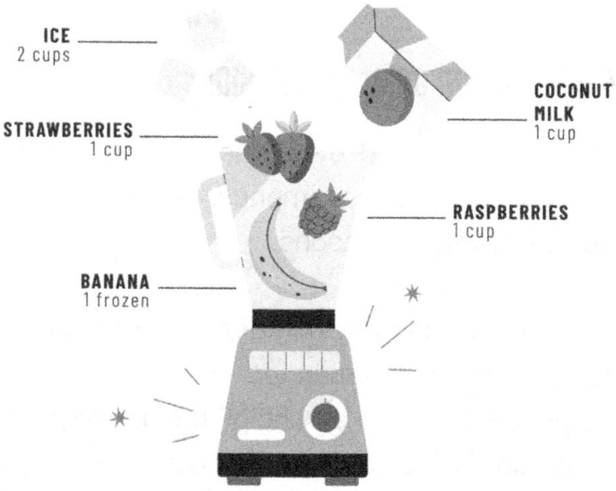

Summer Smoothie Recipe as an Infographic

To understand the difference between infographics and data visualization, we first have to know what they are and how they relate. An infographic can combine text, illustration, data visualization, and icons into a clean, organized progression that enables the audience to understand the message without spending too much time looking at it. They take considerable piles of complex, boring information or data and create something fun, interactive, and user-friendly. A good, strong infographic tells a story. It gives the user all the essential information they might need to make an informed decision.

Well-designed infographics have typographic and eye-catching introductions attracting the reader's attention and providing an incentive to take the time to slow down. Infographics designed to persuade often end with firm conclusions and a call to action to inspire the audience to make a decision.

Infographics are created both for web and print. The difference is that when designed for the web, the layout tends to be formatted for multiple devices and is usually in a vertical, long-scroll format for intuitive scrolling on tablets or phones and horizontal for desktops or laptops. When developed for print, they can be downloaded as an image or printed as a poster. In both cases, how an audience will view the infographic is built into the design.

INFOGRAPHICS VS. DATA VISUALIZATION

One common overlap between infographics and data visualization is that infographics tend to use data visualization in their stories to send a message or incite their audience towards a change. Sometimes designers create an infographic and put a bunch of data visualizations on the page without mapping out the story and creating a powerful, attractive narrative. They include all of the available data related to the topic and not just the most important ones, confusing the audience. A good, well-designed infographic is all about storytelling and not just about making the data look pretty and colorful.

You find infographics used in many places, and with the world becoming more social every second, people spend less and less time reading a document or piece of content that doesn't catch their eye instantly. Some of the significant spaces that infographics have impacted are web and app design, journalism, and even the field of medicine. They have the power to convey detailed stories without making someone feel overloaded with information.

In a world where technology and information are the driving forces of society, the amount of data we need to go through and understand will keep increasing. On the other hand, data visualization is the field that translates lengthy datasets into an easy-to-understand format benefiting the audience. It saves time from looking at incredibly complex spreadsheets by organizing the data into pie charts, bar graphs, scatterplots, flowcharts, etc.

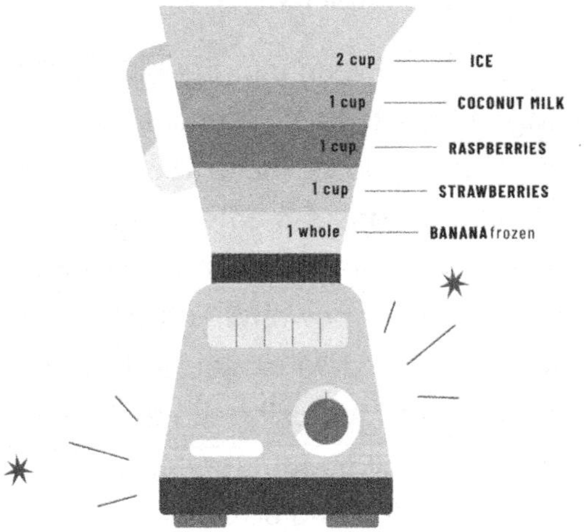

Summer Smoothie Recipe as a Data Visualization

Data visualization is a powerful tool that can translate heaps of technical mumbo-jumbo for the generic user to understand. It helps a viewer detect patterns, changes, and anomalies. The best part about data visualization is that when you look at data organized as a graph or chart, you instantly understand the key points without spending hours leafing through the raw dataset.

Data visualization is a valuable tool in strategic communication. According to developmental molecular biologist John Medina,

> "*Vision is by far our most dominant sense, taking up half of our brain's resources.*"

Moreover, about 40% of our nerve fibers join at the retina, which means that human beings are more capable of processing and storing data if presented in the form of visuals. So if you want to make a more significant impact on the audience, utilize data visualization.

But how do these two relate? How are infographics any different from data visualization when both share lengthy information in an attractive visual to tell a story? Similar to the difference between information design and graphic design, data visualization is under the umbrella of an infographic. It is an element used in making an infographic. For example, instead of writing a descriptive paragraph to describe productivity changes or sales updates over the past quarter, a simple graph can offer data-driven proof to update your boss and save time.

. . .

Data visualizations also use the elements and principles of graphic design but are fully backed by data. It is 100% based on a dataset, whereas infographics are not. It organizes all of the lengthy data within the reader's view, enabling the audience to see trends or patterns without moving the eye around too much or scrolling through multiple pages.

WHY IS DATA VISUALIZATION IMPORTANT?

Data visualization relies on context and is mainly used to compare multiple values of data with each other to provide a specific scenario for the audience. Without context, it is impossible to understand something's value (a piece of data) in the story if it doesn't compare with anything. For example: What if I said you had seven puppies in front of you needing to go for a walk right now? You might think, oh no, that is such a large number. It is chaos waiting to happen. No way can I walk that many furry balls of energy at once. But wait. You don't have any context for comparison in this story. Is that actually an unmanageable number? The values of "seven" and "puppies" are in isolation. Your brain will try to fill in the gaps from past experiences. Do you picture ten tiny adorable furballs? What if I added that they are two-year-old Husky puppies? You could drive a sled with those guys. It would be a completely different story if I said they were two-year-old Daschund puppies, right? You may not be able to drive a sled, but you could probably pull them behind in a buggy quite easily.

You see, it is the designer's duty to provide some context with their information if they want the audience to perceive it in a specific order. If not, the audience will develop their own conclusions based on their individual experiences. Context becomes a potent tool for a designer/creator, especially in the field of data visualization.

> *"The ability to take data—to be able to understand it, to process it, to extract value from it, to visualize it, to communicate it—that's going to be a hugely important skill in the next decades."*
>
> *—HAL VARIAN, CHIEF ECONOMIST AT GOOGLE*

With increased access to instant global connectivity, the volume of information released and consumed is greater than ever, and data visualization helps portray it in a way that everyone understands. It makes sense of complex data in a layman's language for the ordinary person. It grabs attention instantly and helps them retain the presented information for much longer. Without a data visualization, it can be challenging and confusing for a viewer to sift through large quantities of data. Including a visualization makes for more concise and clear presentations.

Some of the reasons why data viz is so popular among companies working with big data are:

- Data visualization strengthens a message's impact by presenting data in a persuasive format. It forms a certain sense of unity across all groups and teams within the organization.

- A data visualization lets you see changes and insights without wasting too much time, enabling stakeholders to make confident decisions based on the data quickly.
- It backs arguments with clear evidence.

All in all, data visualization has the potential to solve tons of business issues, and companies should incorporate it for multiple transformative benefits. Even though more and more data is being thrust toward us every day, we can't digest all of it or use it to our advantage.

But since the whole world is data-driven, the need to consume and digest this data increases, and so does the need for greater data literacy. What is this data literacy? If literacy defines a person's ability to read, write, and understand, in that case, data literacy describes a person's ability to understand, communicate and digest the data visuals sent out into the world in various forms. In today's world, with our extensive knowledge of data and advanced tools, we can offer a much clearer picture of the technical mumbo-jumbo. But to be able to do that, there must be capable designers that translate the data into small chunks of information digestible by the audience. Data visualization has to use design to convert this data into an easy-to-understand format. But it is only possible if you are familiar with the elements and principles of design and know how to use these tools to your advantage.

Data Visualization is important because data-driven communication can turn insights into actions. If communicating information or passing along an insight becomes your goal as you make visuals, you must show the information in a way that draws the attention of the audience, clarifies what the data means, and encourages them to take it further with their actions and thoughts. Without data visualization, it is impossible to communicate data quickly, find patterns, and decipher meanings from data sets, which is why data visualization is one of the most exciting opportunities within design today.

KEY TAKEAWAYS

- Information design is one of the most important ways for a company to portray data in a clear, engaging, and informative manner.
- Information design helps users be efficient in navigating spaces, solving problems, or making decisions.
- There are generally two approaches to information design: infographics or data visualization.
- An infographic combines text, illustration, data visualization, and icons into a clean, organized layout that enables the audience to understand the message quickly.
- Data visualization is the field that creates visuals to translate lengthy datasets into an easy-to-understand format.
- Data visualization uses components that are 100% data-driven.
- Data Visualization is important because data-driven communication can turn insights into actions.

CHAPTER 6
PRINCIPLES OF DESIGN AND WHY THEY RULE THE SCHOOL

THE GENERAL IDEA of this book is to help you become a better designer, data author, and creator. While I can't guarantee that you will become a superstar at design after reading this, I can ensure that you will become a conscious one. At this point in the text, you have an excellent idea of how a person senses a simple visual and how easy it is to control the narrative of your image when you know the inner workings of the mind. But to become a mindful designer, the key is to increase awareness of and focus on incorporating the elements and principles of design using your understanding of human visual perception.

WHAT ARE THE ELEMENTS AND PRINCIPLES OF DESIGN?

The elements of design are the tools and components you use to build a composition. At the same time, the principles of design are a set of rules and schematics that dictates how you use the elements in your layout. The elements and principles combine to create an informative, engaging, and aesthetically pleasing visual to the human eye.

. . .

Every discipline has its own set of rules, requirements, and methods that have evolved to fit the changing world. Similarly, design has its own set of guidelines that it needs to consider to make a visual pleasing to the eye and convey messages efficiently. These rules are known as *principles* in design. These principles are vital as they help us identify and explain excellent design choices or why one visual is better than the other. Moreover, these information design principles are universal and not modified around a specific language or culture.

How will these principles be used in making a data visualization? The purpose of learning about these design principles is that they enable you to understand why and how the human mind experiences pleasure in seeing a visual. For example, a graphic that has a balanced design can have a calming effect. Using hierarchy can direct the viewer's gaze to the most important information on the page. It enables us to create data visualizations by using the tools of layout, color, shape, and text to simplify large statistical data. And according to Tufte, these principles will also enable you to put the "right mark in the right place" in your visual. Combining these elements and principles of design with your knowledge of perception from the previous chapters will empower you to create amazing visuals that are not only fun to look at but also convey a story or other valuable information.

ELEMENTS OF DESIGN

Since we have established the importance of the elements of design, let us look at the attributes that combine to form them. This book will focus on the most impactful design elements in designing data visualizations. They will guide you to produce a visual that is informative, beautiful, and demands the audience's attention.

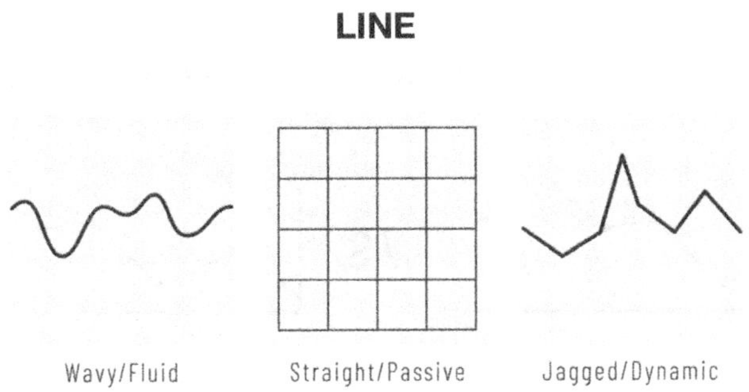

A line is a distance between two points. You must be thinking, "Um, I think I know what a line is. Why am I supposed to go through all the information I already know?" Well, have you thought about how to use it from a design perspective? A line can be vertical, horizontal, diagonal, curved, or zigzag. These different lines have a particular philosophy behind them, and there are extensive studies on their psychological impact on the human brain. The human eye likes to follow lines. Take advantage of this in your design to guide the audience's gaze to the focal point of the visual. If you want attention to move to a specific point, use the continuation of the line to guide it. You can use lines in a multitude of the ways in a data visualization, such as:

Guidelines

Guidelines can be helpful in a visual to help control the viewer's gaze. They make a visual less complicated and easier to read, which ultimately saves time and receives a positive response from the audience. For example, a line graph with many data points can take extra time to read what value a point along the graph correlates to on the vertical or horizontal axis. You almost have to trace your finger from the data point to the axis to make sure you got it right. In a case like this, adding the guidelines improves the visual and makes it easy for the eye to follow.

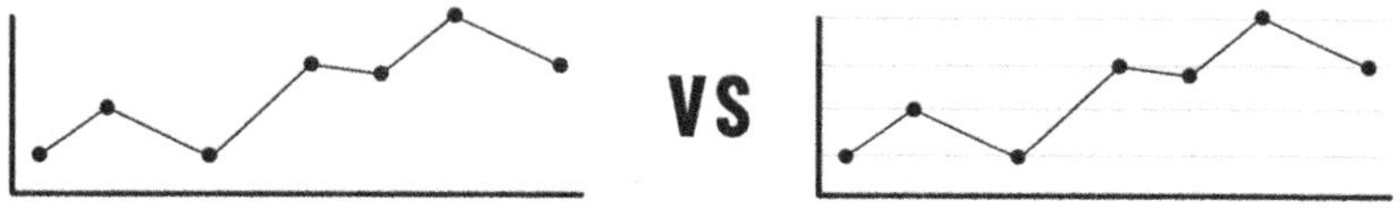

Anchor lines

Anchor lines are crucial in data visualization, especially when you want to highlight a single point in a scatterplot. It helps differentiate the specific data point from the others and supports the comparison between the most important data. They can be used in column, bar, and line charts to help provide more visual context.

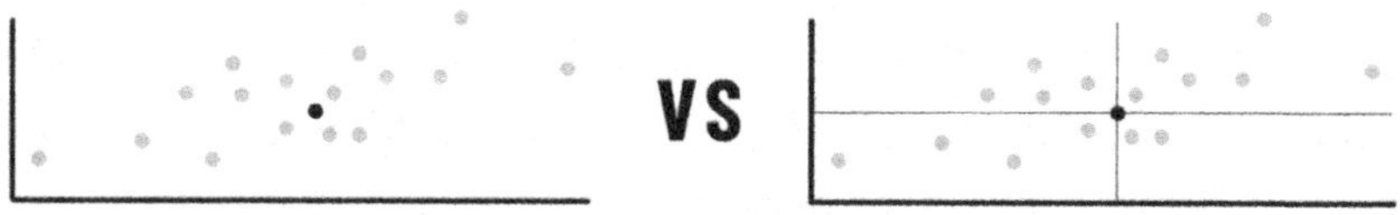

Trend lines

Like anchor lines, trend lines help distinguish a pattern in a scatterplot or line graph that helps display the relationship trend between different points at a glance. It also helps define the positive or negative relationship with the slope of the line. It is a quick and effective tool to provide instant information.

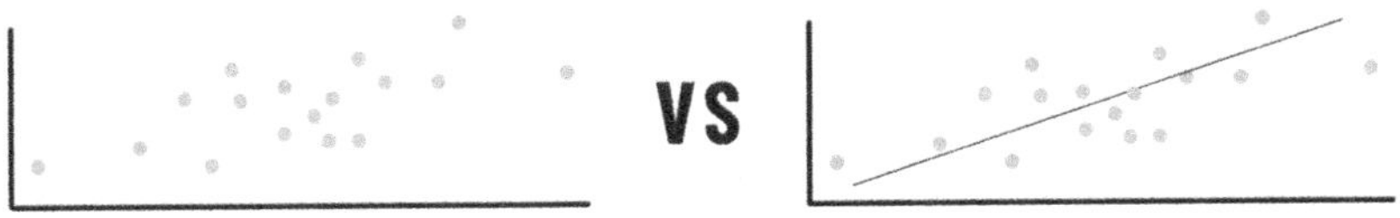

Shaded region adjacent to the line

These work exceptionally well in charts when you want to easily bring attention to a grouped part of the visual. For example: In a bar chart showing positive and negative values, the negative value region could be shaded with a semi-transparent color to help the user quickly interpret and distinguish the data between the alternating values.

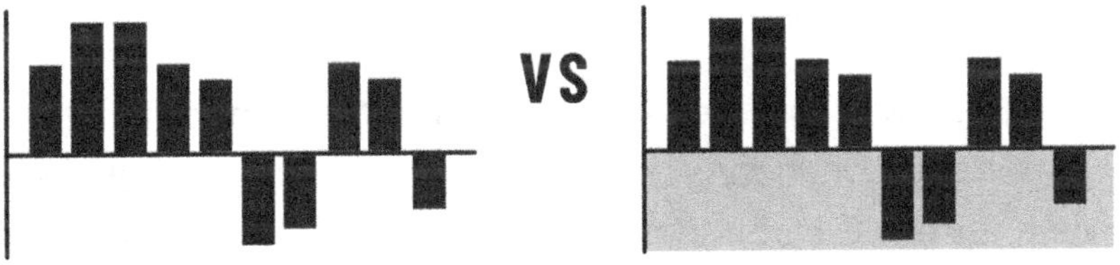

Gridlines

Gridlines are a fundamental and subtle element of design that can add great value to your visual. They help break down how you can organize your layout and can also be used to create a frame for your data within a blank space. Depending on the number of other elements in your design, it may also be helpful to remove gridlines for a cleaner presentation. Later, when we discuss layout and composition best practices, I will expand on different methods to use grids and gridlines to benefit your visuals.

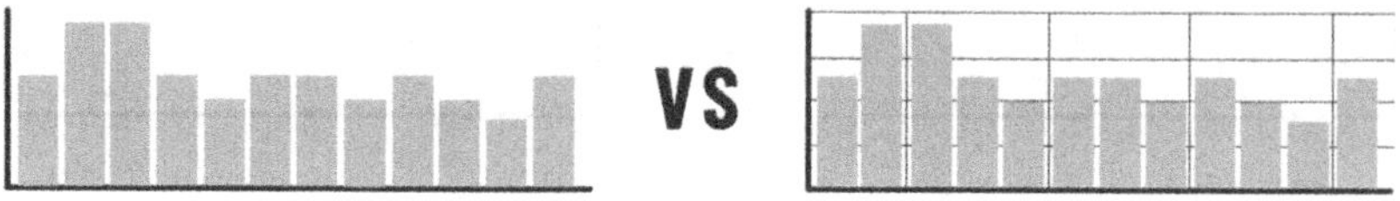

SHAPE

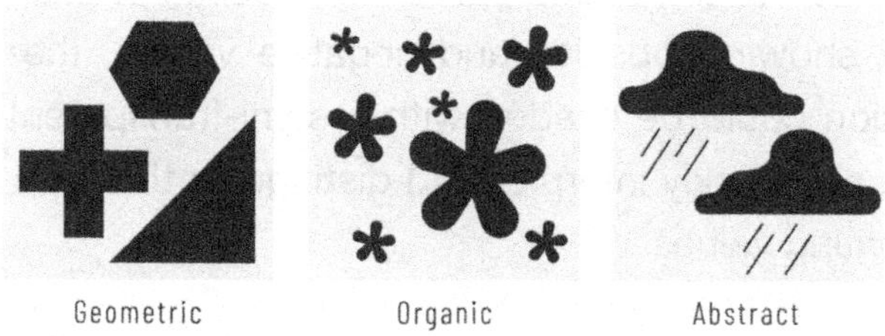

A shape, as you very well know, refers to the form of an object or its external boundary or outline. A shape is a 2-D element, while a form is 3-D. Shapes are one of the first things we teach kids and something that gets internalized into long-term memory. Do you remember trying to fit a square block into a round hole? So when our eye sees a shape, it refers back to the LTM and tries to make sense of it by comparing it to a previous memory.

Shape is tricky in design because the way we look at or perceive a shape can change based on the shapes surrounding it. This is most apparent when looking at the Gestalt principle of figure/ground, also known as positive/negative space. With this principle, the positive shape is known as "figure," and the negative space is "ground". The basic idea of figure/ground is that the audience's eye tends to separate whole figures from their background. What they choose to see as foreground and background is based on their pre-existing knowledge. It is one of the first and most automatic things people do when they come across any composition.

FedEx

Once you see the arrow between the "e" and the "x", can you ever not see it again?

In such relationships, the arrangement of the different shapes makes one of them fade into the background while the other shape pops out. If it is obvious these shapes are either figure or ground (foreground or background), then it is thought of as a stable relationship. If the different objects/shapes can switch in the mind and appear either in the foreground or the background, it is a multi-stability relationship. A great example would be the logo shown above. Since our brains can only focus on one at a time, the shape you look at for longer appears to stay in the foreground. Focus on the arrow in the logo, and it will look dominant. Focus on the words, and the arrow fades away.

TYPE

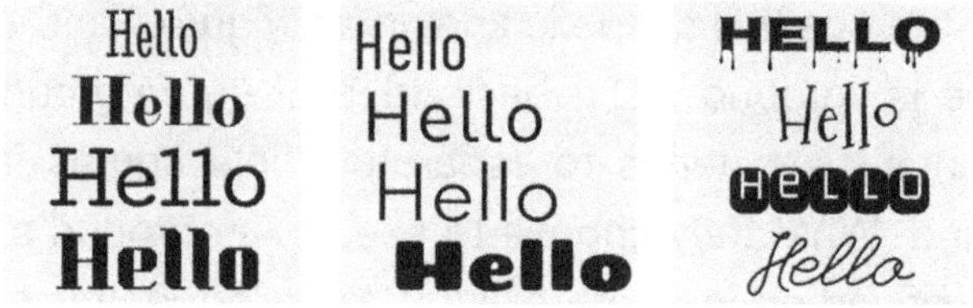

Serif Sans Serif Decorative

Text is an integral part of a visual and a practical element for storytelling through design. Typography refers to arranging typefaces in various fonts, sizes, and spacing. You can use typography to portray a mood or theme or align with an environment. For example, sans serif fonts tend to be more appropriate for business presentations, while serif fonts can be associated with literary or classical themes. If you make a viz for your portfolio, try experimenting with font choice to add an extra layer of interest and story.

Graphic designers tend to use typography as a potent tool in creating expressive visuals that impact the viewer. Font type, size, color, and weight can be played with to create hierarchy and visual groupings. Many brands have distinctive typography that becomes their sole identity. But even typography comes with an attached rule book. The goal of playing around with different possibilities of a text is to create contrast or highlight the specific part of the visual. Words can lose their significance if they are catchy but have poor type treatment.

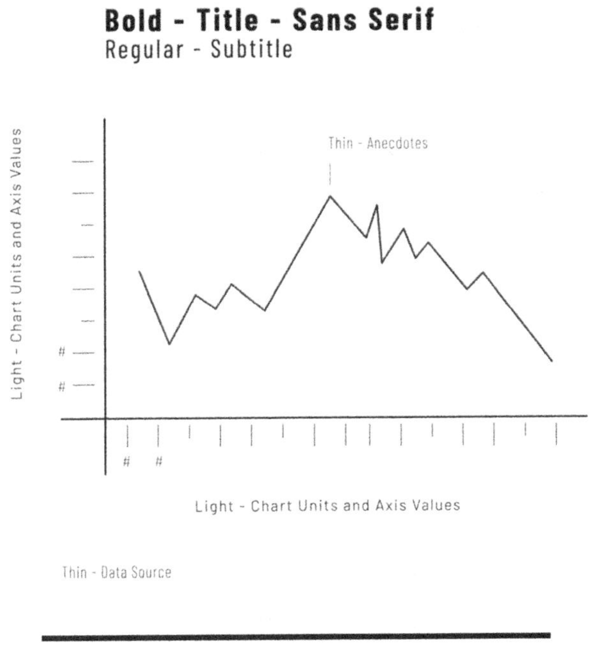

Create visual hierarchy by adjusting font weights.

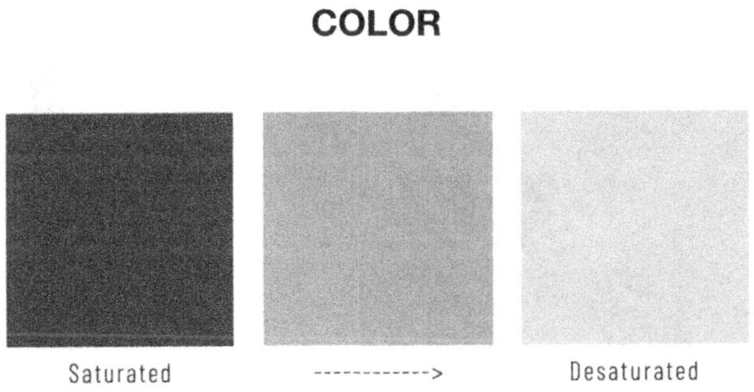

Color has to be one of the most studied elements in psychology. It is a potent element as it subconsciously connects with the user's mood, emotions, and feelings. Every color has a different characteristic or relationship to the viewer. Using the right color in the right amounts can impact how well the audience receives your visual. Have you ever wondered why one person lights up at the sight of yellow while another surrounds themselves with the color purple? Colors are strongly related to feelings, emotions, and memory. For example, people consider blue a calming color, while red relates to danger, heat, or love. Further study of basic color theory and color pairings can help with consistency and better color choices in your designs.

Let's say you are making a visual around a summer theme. For such a project, you could pull colors from a summery photo to create a beautiful color palette. The colors for your data could be bright and glowy to make the user think of that season. If you are designing a visual for a brand or about a specific brand, an obvious choice would be to use brand-appropriate colors to identify them as necessary to the story.

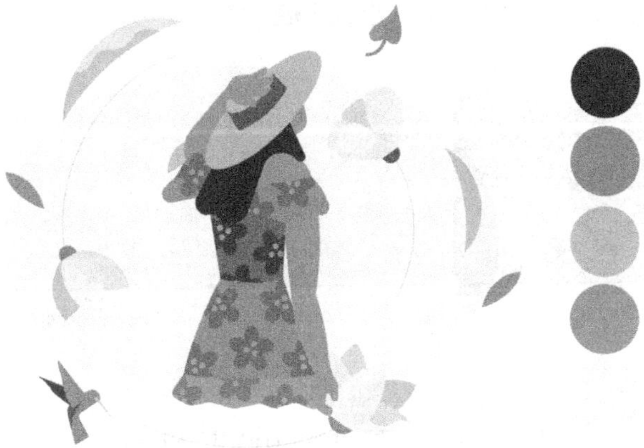

If you are stuck on choosing colors, try sampling a range from a favourite photo or illustration.

While it is good to use color in your design to attract attention, you must also remember to use color responsibly. Using too many colors or choosing colors that don't make sense can make the visual too chaotic and busy to look at, discouraging users. You want to bring them in, not push them away.

PRINCIPLES OF DESIGN

Principles of design are the basic guidelines for mixing and matching a visual's design elements. The elements are the *what*, and the principles are the *how*. The better your understanding of these principles, the better equipped you are for the design phase. Here are some basic principles of design you can use in your next project that every designer, whether rookie or professional, should know.

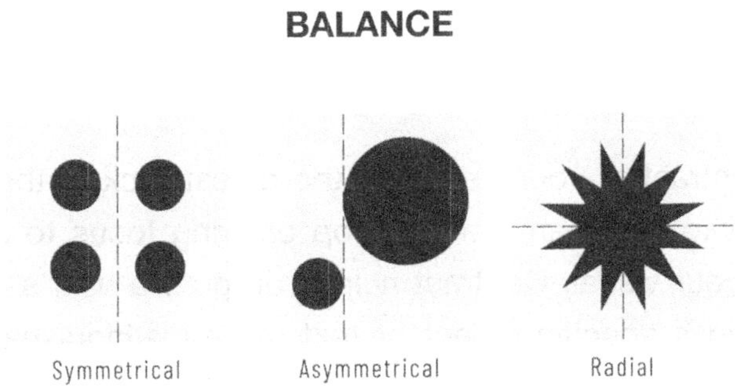

There are several types of balance in a design: symmetrical, asymmetrical, and radial. Balance is one of the top principles of design that you must not forget. It refers to the visual weight of your elements based on the distribution of objects, color, shape, size, space, etc. Balance gives a design structure, stability, and a feeling of form. But that doesn't mean that balance makes a composition rigid and that everything must be symmetrical. It helps you navigate the audience's gaze in a certain way that enables them to look at and perceive your design quickly and clearly. Balance carries associations such as harmony, order, rationality, and peace because the eye of the audience does not have to move around back and forth between materials to understand a design. Balance also helps you avoid any kind of chaos and makes your design aesthetically pleasing.

CONTRAST

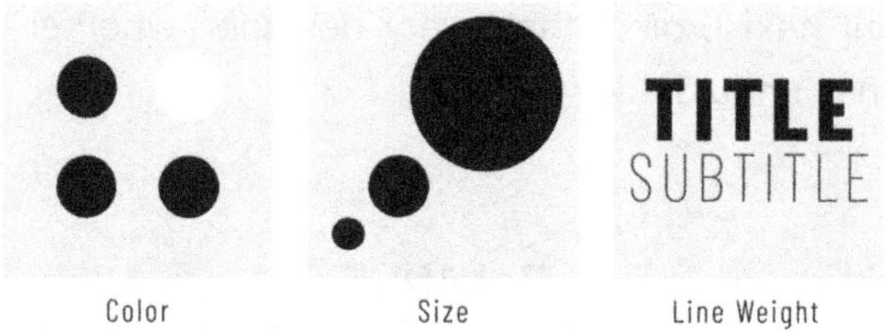

Using contrast in your design is the oldest trick in the book if you want to make an element pop or bring focus to a certain point in your visual. Contrast helps you grab a user's attention by making a specific object or text more distinguishable than others. You can achieve this by placing two design elements in opposition to each other.

Contrast can be created in many different ways. For example, you can pair light with dark, small with large, or thick with thin. The most common ways to apply contrast in a data viz are using contrast colors for your chart elements, pairing thick text with thin text (as with a title and subtitle), and utilizing size to show the comparative value of data points. In addition, these visual differences can help improve the readability of the design by engaging and speaking directly to the viewer's sensory memory.

PATTERN / REPETITION

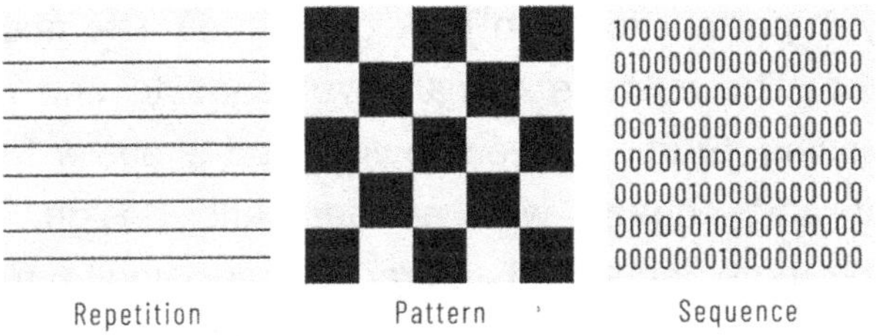

A pattern forms when an element, object, or movement repeats over and over again. It creates visual groupings. Repetition is a terrific way of ensuring symmetry and beauty in your design. Repeating elements such as fonts, colors, shapes, or text can help bring together your visual in a holistic manner. It can make your design beautiful and inviting and encourage viewers to spend more time looking.

WHY IS THIS KNOWLEDGE IMPORTANT?

After learning the science behind visual perception and design, you can now gauge how people think and react to visuals. Unfortunately, your visual cannot target and resonate with everyone. Therefore, focus on designing for a specific target audience. Consider their reaction to the visual. This will give you a stronger idea of the narrative to propose. Having an in-depth knowledge of this will enable you to make designs according to the specific needs of your boss or client. Whether for a presentation or portfolio, you will completely control how the audience understands and responds to that visual.

Moreover, this knowledge can give you greater confidence in creating informative visuals that tell a story and look beautiful. But if your visual communicates well and tells the data's story, is it really important for it to look good too? The answer is a definite yes. The average human eye tends to find attractive things pleasing. Studies have shown that it is easier for people to accept and engage with an aesthetic design. Pleasing designs promote increased interaction and incite a positive feeling in the user, promoting creative thinking. Not only that, but people find it easier to forgive faults if they are present in an aesthetic design than they would otherwise. These are great incentives to experiment with the elements and principles of design to make beautiful and practical visualizations.

This knowledge of the principles and elements of design will steer you through creating those stunning visuals. It will give you the confidence to go ahead with your creative ideas and help you not feel intimidated by the work of your peers. Instead, you are in a position to identify what makes their designs beautiful and can incorporate it into your own work. This knowledge also makes you design-conscious and expands your design mind. You can gauge a bad design from a good one by taking a small look at it! You are armed with a new batch of tools and guides to consider how you can revamp current work and apply this knowledge to future projects. This information will help you take a step forward in your career as a creator or data visualizer. And all of the things we have learned up till now are timeless and will remain relevant to your field whether you design visuals in 2022 or 2050.

KEY TAKEAWAYS

- The elements of design are tools that you use to develop a composition.
- The principles of design are a set of rules and schematics that a person may follow to use the design elements to create a visual narrative.
- Some elements of design are line, shape, size, type, and color.
- Some principles of design are balance, contrast, pattern, and repetition.
- Having an in-depth knowledge of this can give you greater confidence in creating informative visuals that tell a story while looking beautiful.

CHAPTER 7
EXTRACURRICULAR: LAYOUT AND COMPOSITION

IN THE LAST FEW CHAPTERS, we talked about the components of design, how they are perceived, how they react to emotions, and how you can use them to convey a powerful message or a story. But this is where things get both: tricky and fun. This chapter will discuss everything you need to know about layout and composition in design. What are they? How do they work? Now, you can get clever with implementing all of your new knowledge into your work.

What is so important about a well-designed layout? A layout or composition is how you arrange your visual elements within a space. It creates structure and organizes the elements within the area. As you go through your pre-planning, sketch out an initial strategy for your layout. Consider where components such as labels, annotations, or titles might fit. What graph type do you want to use, and how will that affect the other elements? It's ok if you have to make changes later but start with at least a rough idea. Any design requires a balance between the visual arrangement and its message or intention. This balance is only possible if you understand the rules behind a layout and use it effectively to narrate your story or explain your design.

PRINCIPLES OF LAYOUT AND COMPOSITION FOR AN EFFECTIVE DESIGN

You know the layout is a critical part of visual design, and you must be wondering, "why have we waited until Chapter 7 to learn about it?" Layout is significantly interrelated to the principles and elements of design, as we learned in the previous chapters, but you need to have an overview of these principles before putting them all together.

The layout is about connecting the dots. It helps the brain understand the relationship between components of a visual and how to emphasize the most important details. One way to explain this is with the alphabet. Individual letters usually don't make any sense when they are floating alone. When these letters group together as a word, you begin to see a story. When these words are organized in a phrase or sentence, you start to get the big picture and feel an emotional connection. In this example, letters are the elements, the words are the principles, and the sentence becomes the layout or composition.

GESTALT

A well-designed layout is not only a treat for the eyes but also the mind. Gestalt, a German word that means form or shape, refers to the theories of exploring perception and behavior when people look at something or process it as a whole system. "The whole is greater than the sum of its parts" is a saying by Kurt

Koffka and is one of Gestalt's most understood concepts. Multiple Gestalt principles are related to design and data visualizations. They are:

Figure/Ground

Figure and ground refer to the foreground and background of a visual or an image. I have previously mentioned this principle in the last chapter. Figure/ground or positive/negative space is the most creative way to form your layout. It creates visual attraction and meaning while encouraging the audience to look at a graphic for a minute longer. In addition, it creates either a stable or dynamic relationship between elements or shapes in a layout.

Continuation

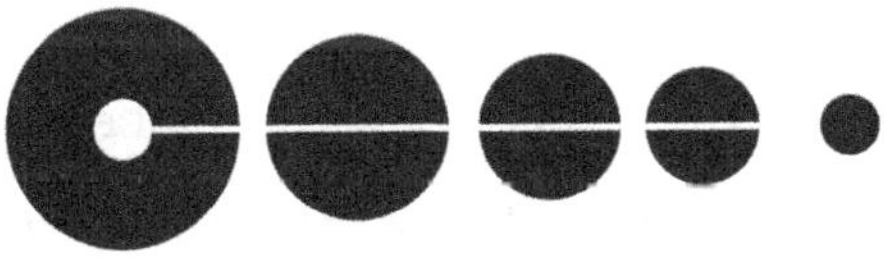

Continuation implies the act of orienting the eye of the audience in a way that moves throughout the visual fluidly. You can use this to help control and improve the eye movement of the viewer.

. . .

Similarity

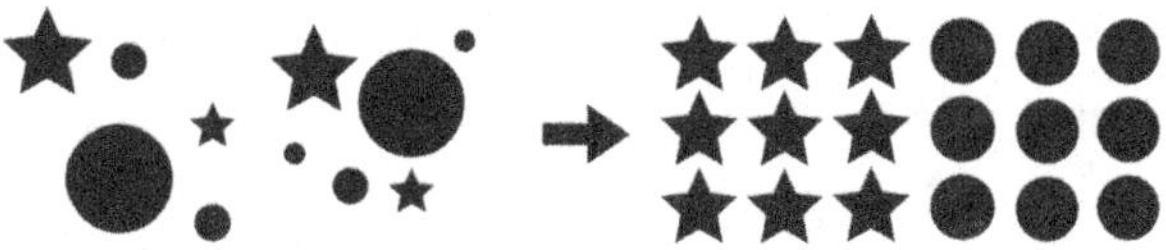

Similarity means a grouping of comparable things together. We learned that chunks of information are more manageable for the mind to understand and comprehend. Similarity uses visual attributes to group objects and reduces cognitive load. You can create similarity by grouping objects of similar size, color, shape, or alignment. It can help you keep your audience's attention longer without confusing them.

Symmetry

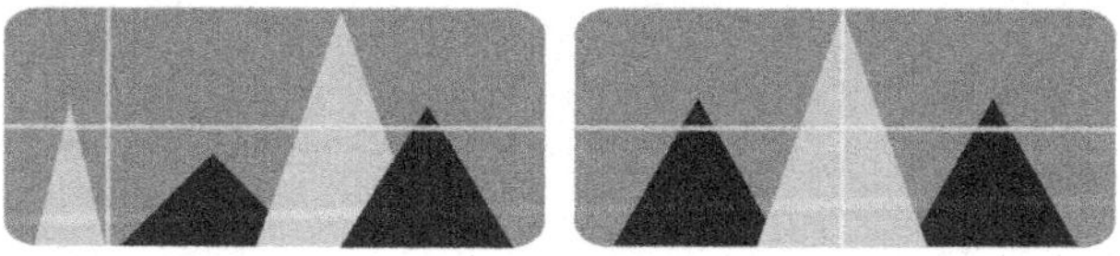

Symmetry refers to recognition and balance. You can have equally spaced groupings that are symmetrical, or groupings weighted to one side that creates asymmetry. Symmetry is a great tool to use in your layout design as it provides balance and harmony for the eyes. It helps the mind recognize patterns and enables the user to interact with the visual more. It also makes it easier to understand the material as everything has a certain order, placement, and sequence.

. . .

Common Fate

Common fate suggests that items that are associated must travel and stick together. When images or objects have similar visual characteristics, grouping them makes it easier for the mind to perceive and understand the material clearly and effectively. It can generate a sense of motion and add energy to a visual.

Closure

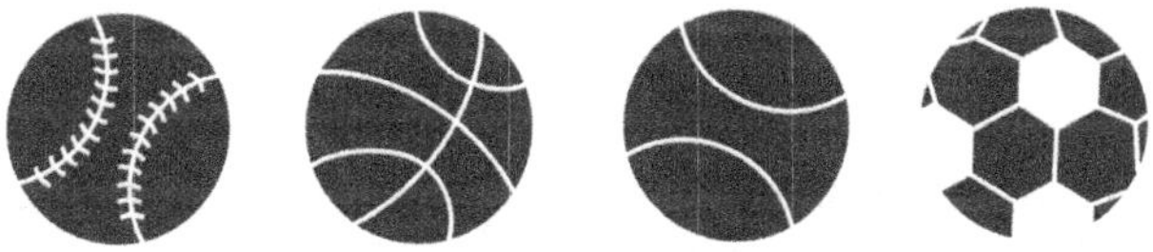

When applying closure in a design, you take advantage of how the brain will fill in the gaps of familiar shapes. When our brain looks at a form with gaps, it completes the picture and processes it as a whole.

These principles are not just for you to go over and weakly remember when you design. Utilizing them in your design to the fullest is what's important. Consider all of the possibilities of these principles in your viz to engage your audience and make them go from passive observers to active, interactive users. Breaking down complex pictures into simpler shapes and forms can change how an audience perceives your design. Complex ideas do not always have to come off as busy, chaotic visuals. But instead, by using the principles of design and layout smartly, you can ensure that a visual is clear, concise, and interactive to your audience.

Think of it this way, what is the last 'good' design you remember seeing? What made it good? What made it stick in your brain among the many different ones you encounter daily? Whenever we, as users, look at a visual and connect with it, we are likely to process it in our long-term memory, as we studied in Chapter 3. But how do you actively incorporate these principles to connect your visualization to your audience? How can you use the Gestalt theories of perception to make your viewer an active participant in your design?

A design or a visual does not have to give it all away. Instead, using an image or form that the viewer can put together themselves ensures they actively partake in the design. Looking at a visual and having to work it out activates the problem-solving part of your audience's brain and encourages them to spend more time with your design. Once they make the connection and have a little aha moment, a feeling of pride emerges as they see themselves as the problem-solver. This is a terrific strategy to bring out a positive emotion in your audience. Experimenting with principles such as figure/ground, closure, or symmetry can encourage curiosity. Your audience will feel clever and feed your design to their long-term memory.

ASPECTS OF LAYOUT AND COMPOSITION

If you have been intrigued to try and figure out layout and composition for a better, more effective design, you will need to understand some components to get you started:

GRID

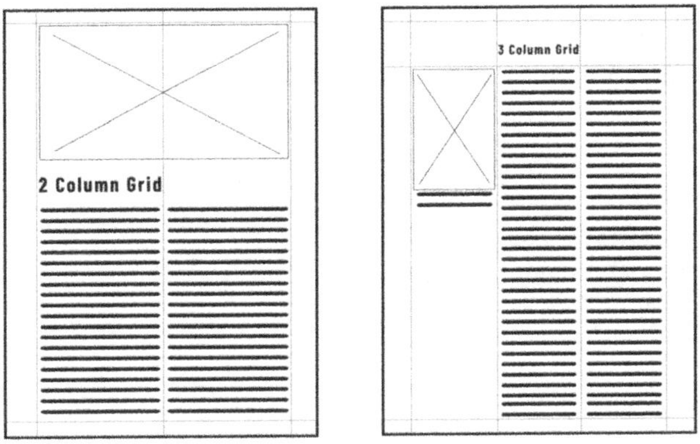

A grid is one aspect that correlates to alignment, symmetry, and balance. A grid is a system of lines making up columns and rows that divide a 2D plane into smaller, measured sections to make image and text placement easier. There can be numerous grid types with different amounts of these columns and rows, but you must choose what makes sense for your design and the content you will fit within it. Grids are useful because when you place text and images systematically within, the priorities stand out clearly, making it easy for the viewer to navigate the layout. The main benefit of a grid is that it maintains the flow of the design, which creates a better user experience.

. . .

The most common grid system used in layout is the multiple-column grid. It is the simplest to navigate and forms the perfect symmetry for the human eye. A column grid is a row of vertical sections with a space made up of columns. These vertical spaces become section breaks for the eye and create a frame for the entire layout. This way of organizing has the same effect on the human eye as the golden ratio and is simple and easy to incorporate into your design. Take a look at a magazine or newspaper and count the number of columns used to organize the text. It is standard to see 2 to 3 vertical columns in magazines and even more for newspapers.

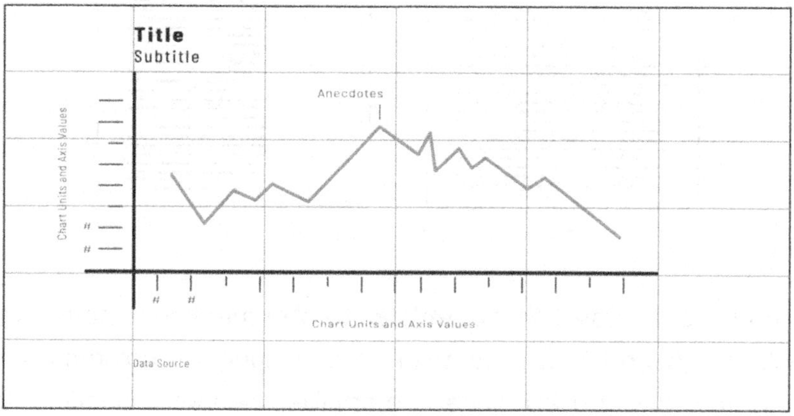

A 12-section grid laid over our previous example of hierarchy via typography.

You can experiment with how many sections to use, but try starting with 12 (6 columns and 6 rows). These sections can easily break down to smaller numbers, such as 6 or 3. If your data story reads from left to right, you can think of your columns as acts in a play. What happens in the first act in column 1? Is there a hero's journey in column 2 that leads to a climax in column 3? This can lead to a final resolution in the last column. But what if you don't want to follow a rigid 12-section grid for your layout? You can go off-grid and add variety to the design. Breaking off the grid can be challenging to get right, especially for beginners, as it requires a lot of balancing and experimentation. Keep your audience in the forefront of your mind during planning. It is all about the viewer at the end of the day. Nonetheless, it is not impossible to do so, and as with all creative endeavors, it is fun to break the rules a little. Don't forget that the primary and most important role of a grid is to make alignment, containment, and organization more straightforward and fast.

> *"Just as maps, roads, and buildings demand a virtual geography, all data must have a spatial home if they are to be seen."*
>
> - RJ ANDREWS - INFO WE TRUST

SPACE

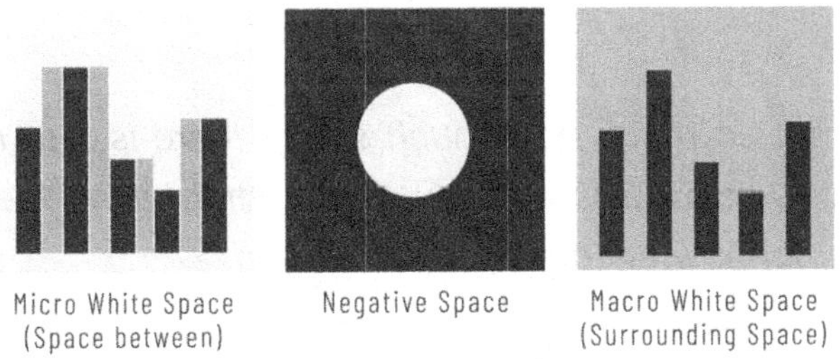

Space is probably one of the most challenging yet effective ways to design an interactive layout. In a layout, the space between your graphical elements is known as white space or negative space. Think of this space as punctuation among all the different elements on a layout. Previously it was thought of as wasted space, just something to be filled. But the more people spent time with the blank space, the more possibilities arose for its use in interactive design. Learning how to use this negative and positive space to your advantage can be a real challenge.

Space between visual elements allows the viewer's eyes to rest, giving them a chance to absorb and collect all of the information without getting too swamped. It helps define your groupings and can emphasize what is not included just as much as what is. It can create clarity, sequence, and shape and give the eyes a little room to breathe between elements. Paying attention to the use of space is crucial in a layout as it draws attention to its surrounding components. Use negative space as a design element or tool to highlight key parts of your visual. Space is also fundamental in a layout because it helps group similar elements together into chunks. And as we know by now, pieces of similar information are more manageable for the mind to perceive, and ultimately, the user will better comprehend your design.

> *"It is not so much how much space there is, but rather how it is used. It is not how much information there is, but rather how effectively it is arranged."*
>
> – EDWARD TUFTE - ENVISIONING INFORMATION

ALIGNMENT

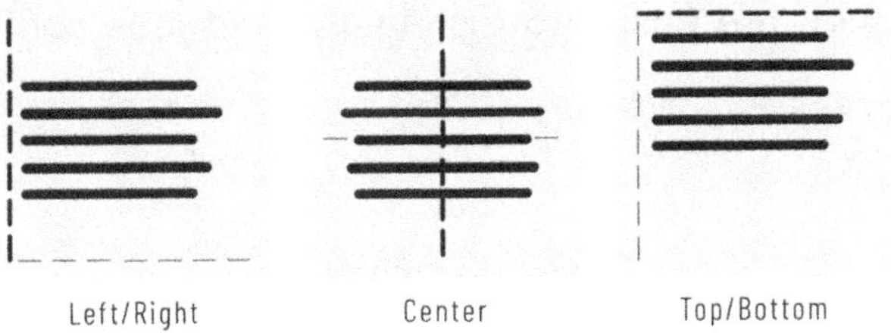

Alignment is another aspect to consider while designing your layout. Following a specific alignment of objects helps you decide where to place things (such as images and text) while also creating a certain harmony between these elements within a layout. There are different types of alignment. You can place elements along a left or right edge, or stack them along a center line. For example, think about how you might place text in the cell of a spreadsheet. You might choose to align all the content of the same column so that it visually groups together.

PROXIMITY

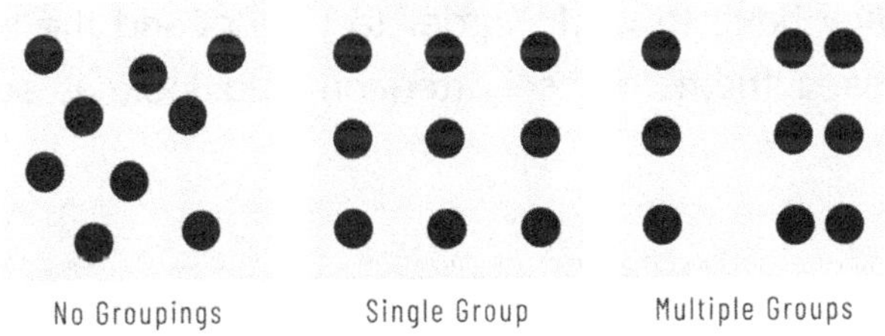

Proximity refers to the position of objects at a certain distance to convey a specific meaning or message. It can differentiate between two or more things or create visual groups. Proximity also defines relationships between elements without words. For example, if we see two components near each other, we assume they are related or similar. Unrelated elements, when spaced apart, are easier to distinguish and less confusing for the user.

HIERARCHY

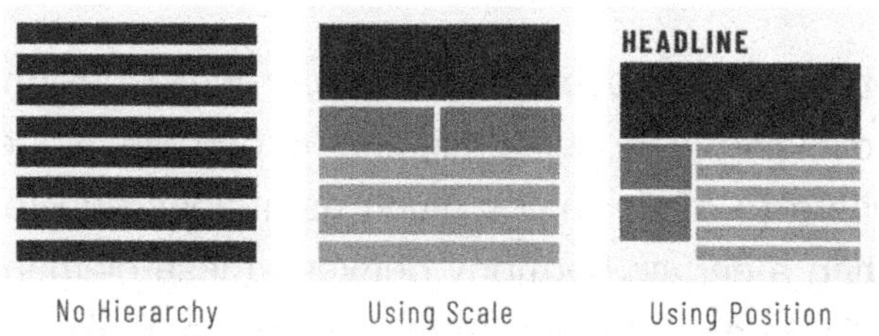

A proper visual hierarchy allows your audience to scan the page logically before focusing on the most important message that your designed visual portrays. It enables them to spend less time looking at a design and still attain the maximum information possible from it. Poor hierarchy can make a layout ineffective and frustrate your viewer. If details are mixed up and difficult to navigate, no one will want to interact with such a design. They have to work harder to understand the message, which causes them to lose attention and look at something else.

As we covered in Chapter 3 of this book, pre-attentive attributes are the attributes that get detected automatically and without a conscious effort by a human eye. This automatic and unconscious type of processing that occurs when we see an object happens at a very high speed and tends to detect only a specific set of visual attributes. These attributes signal the audience where to direct their gaze and are used to catch the viewer's attention before they can focus and look deeper at your visual. By understanding the specifics and the science behind these pre-attentive attributes, you can organize a layout that complements the human eye and mind. Use them sparingly and purposively to ensure an uncluttered, good design that speaks to your audience in the most interactive and personal way.

KEY TAKEAWAYS

- A layout focuses on the proper arrangement of text and visuals within a design.
- Layout and composition are about connecting the dots. They help the brain understand the relationship between components of a visual and how to emphasize the most important details.
- Gestalt theories refer to the exploration of human perception and behavior when people look at something or perceive it in their minds.
- Some principles of Gestalt include positive and negative space, continuation, repetition, similarity, closure, symmetry, common fate, and prägnanz.
- Aspects of layout and composition are grids, space, alignment, proximity, and hierarchy.

THE UNBELIEVABLY EASY WAY TO GET YOUR GOOD KARMA FOR THE DAY

You, yes, you (you look great today, by the way), can (right now) make an impact on someone else's life that could shape the course of their future. That sounds pretty intense, I know, but it is still very possible. Words have power, and your voiced opinion and experience can be the tipping point that inspires someone to take action for themselves, just as you have.

Someone else is out there right now, less experienced, full of eagerness to gain new skills and expand their creativity, but unsure how to get started. This is where you come in...

The only way I can help others and provide value with my books is by reaching them. The absolute best way to do that is through people sharing their honest thoughts in reviews. Reviews are incredibly important to us authors to not only have a better chance to reach readers but to hear your feedback and keep improving.

Your review will help...

...one more student level-up their school projects.

...one more data scientist really impress their boss.

...one more non-profit raise funds through even more engaging presentations.

...create a data-driven future that is accessible and empathetic.

To make that happen, all you have to do (this can take less than 60 seconds) is leave a review. If this book has been of any value so far, would you kindly take a brief moment and leave an honest review? I read every one and incorporate your feedback into future books. Thank you so very much.

To leave a review, please scan the QR code below

Scroll till you find the book in your recent purchases and fill out the form. You get extra good karma points for adding a photo or video to the review.

PART THREE

APPLY YOUR DESIGN MIND TO DATA VISUALIZATION

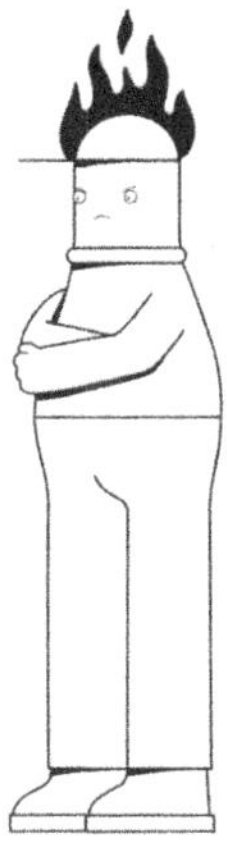

"Information is only useful when it can be understood."

MURIEL COOPER

CHAPTER 8
PRE-FLIGHT CHECKLIST

FLYING a plane has to be one of the most meticulous jobs in which many people's lives (including your own) depend on you. If you are taking off for a flight, you must follow a series of checklists known as a pre-flight checklist to ensure everything with the plane is in order and working perfectly to steer clear of any mishaps. A pre-flight checklist is necessary regardless of how many times you have flown before; you cannot rely on the feeble human mind to remember every little detail.

Similarly, when you start off with a design or a data visualization, it is helpful if you can manage to map out everything you want to position, narrate, or explain in your visual. It would be best to have a specific story that you want to explain to your audience. Therefore, before you start your data visualization, you must have answers to a couple of questions. These questions will be of the nature: Who is your audience? What story, information, or agenda do you want them to learn through your data?

WHAT COULD BE ON YOUR CHECKLIST?

Include Context

So, what should a design checklist include before you start the process of data visualization? The first thing we ought to clarify here is that data visualization is all about context. Context is vital for 'setting the scene' for your viewers or audience to understand your point of view and the message you are trying to convey. You must always provide the context instead of assuming the audience will have a preconceived understanding. This way, you ensure that your audience absorbs your design and message rather than it just bouncing off. Remember that if we see a number or value by itself, our mind doesn't know how to interpret it, and the viewer ends up confused.

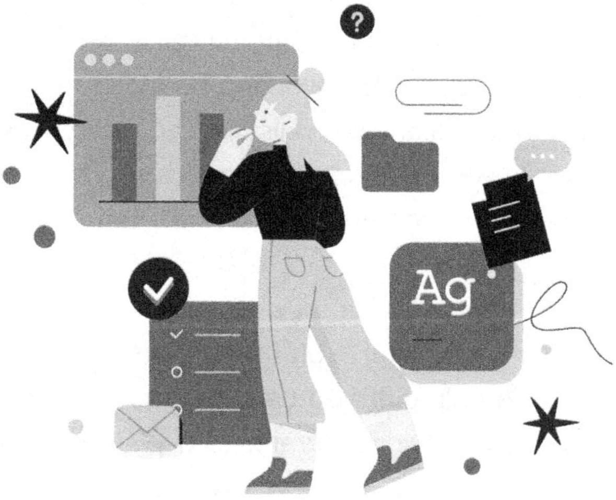

Since the brain has difficulty understanding a word or letter when it stands in isolation, it doesn't know how to categorize it. Whether the data point is big or small, good or bad, increasing or decreasing, we don't know until we provide the context and compare it with other existing figures. Therefore, if you fail to provide context for your audience, they will most likely make one up, and the possibilities would be endless. The intended message will get lost in confusion without sufficient context.

But what is sufficient context? It is up to you to decide what information is crucial to show and what you can leave out of the visualization. You must be able to create a balance bias. Balancing out bias means finding harmony between communicating your message, representing important data accurately, and minimizing the potential bias in your design or data visualization.

Know Your Audience

Ensure that your data visualization speaks and connects to your audience by knowing who your audience is. Starting with this knowledge will help dictate how you design. For example, a chart designed for corporate workers will not be appropriate for high school students with different data literacy, etc. Therefore, your audience should be one of your first concerns.

The more specific and detailed you are about your audience, the better you will be able to design for them and ensure successful communication. If you try to communicate with too broad a scope of people, you will lose the effectiveness of your communication. You are more likely to connect better if you narrow your audience and target a specific viewer for your visual to speak to. The feeling of disconnection between you and your audience can make a great design unsuccessful. Moreover, determining a target audience will make your design more sensitive, appealing, and informative. Focusing on their needs and interests will teach you how different groups of audiences tend to perceive data.

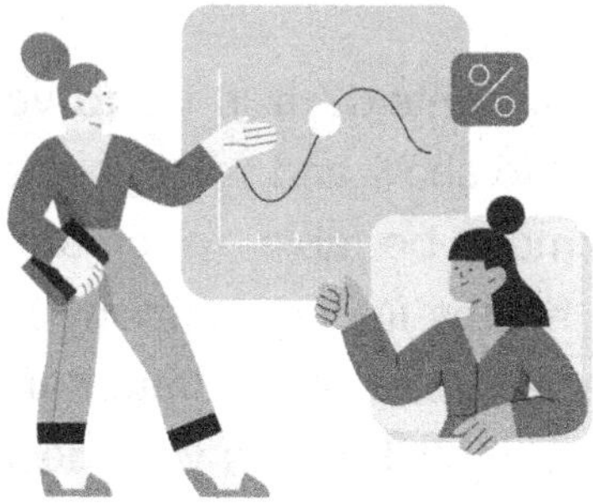

Consider how familiar your audience is with data visualizations and how you explain your data. The better you know your audience, the more you will be able to design a visual that resonates with them. If you are presenting to a group of people who don't understand statistical terms and technical language, you might need to skip them and explain your context in layman's terms. Also, think about how much time your audience has to interact with your graphic. If the audience has little time, patience, and interest, simple charts and visuals are the way to go. But if your audience has lots of time and interest, you can get away with designing interactive charts and detailed visuals.

When you design with a specific audience in mind, you hone in on their likes, dislikes, and way of perceiving things. Then, you calculate how much time your audience has to look at your visualization, what information they require, what they already have, and the information they expect you to provide. This is needed to create something that people relate to and respond to.

. . .

Your next consideration should be what you expect your audience to feel, learn, or do from your visual. Each data visualization has an intention behind it. If you don't have any story, moral, or information to convey to your audience, it is better if you don't design anything at all. All information you put into your visual must deserve its place and engage with the audience's mind.

Establish Your Narrative

Now that you understand the importance of knowing 'who' you are designing for, the next crucial step would be to know the 'what' you are trying to deliver to your audience. It is essential to be aware of your main idea or narrative before starting your design. Knowing this will help you figure out how to design to convey maximum information to the audience in the best way possible. In addition, knowing what you want to share will give you an idea of how to design it. For example, do you want to highlight a comparison or trend, change over time, or even correlation? You will determine this after deciding your message or main idea.

Now that you established the 'who' and 'what' of your visual, move to the 'how.' How do you set up a data story with appropriate context? A good narrative is all about compelling storytelling. Many great designs follow three classic steps: introduction, key message, and conclusion.

1. **Introduction/Foundation**: This section of your data story focuses on forming a foundation for the information

you will share with your audience. This part helps lay the groundwork for your visualization to be understood and perceived by the viewer.

2. The Main Event or the Ah-ha Moment: The ah-ha moment is the main event of the data visualization. For the design or visualization to be incredible, it must incite excitement or a moment of understanding in the audience. And that will only happen if the visualization contains some new, previously unknown piece of information.

3. Conclusion/Call-to-Action: There must be some closure at the end to round out your data story, just like a powerful ending of a speech. If you design a visual to incite emotion in the audience, this final section of the presentation must excite people and urge them to act on their feelings. Should they visit a website, sign a petition, strive for change, make a business decision, or donate to a cause? The visual should leave the audience with the knowledge of 'what' to do with the little nugget of information provided to them by you, the designer.

Once you have a basic sketch of what you want to convey and the key takeaways your readers should get from your visual, combine it with the appropriate graph type that brings your ideas across in the most effective way.

HOW TO CHOOSE THE RIGHT DATA FOR YOUR VISUALIZATION

If you compare your data visualization to a museum exhibition, your data would be all the pieces that make up the collection, including what wasn't displayed. All the potential art pieces are gathered, sorted, and specifically chosen or not chosen for an exhibition. If sufficient data is absent from a visual, it becomes challenging for the viewer to build an accurate picture of the data story. It doesn't mean that you put all of your data into one visual and make it, again, impossible to understand. Instead, learn to choose what is most relevant from your data set. Curate your dataset and determine what makes sense for your 'exhibition.' Sometimes, this could mean that you have to manually add data so you don't miss out on intriguing variables to visualize.

In the book *Data Sketches*, Nadieh Bremer created a visualization about all the vacations she had taken since she was born. Since it was personal data, she had to collect everything manually, but it also provided an opportunity to make something fun and interesting. The obvious data points to use are dates, destinations, and length of the trips, but Nadieh added extra details such as who she traveled with, the purpose of the trips, and how much she enjoyed them. Expanding out in this way creates a clearer picture of your data story as well as more opportunities to connect with your audience emotionally. Learn to sort through your data, eliminate clutter and add the data that makes the most sense for your idea and visualization. Be the curator of your own exhibition!

To know what the correct data is, you start with understanding the multiple types of data that exist and how they relate to one another. Data comes in many types, but if you are designing data visualizations, there are two to remember: quantitative and qualitative. Quantitative data refers to values that can be measured or calculated numerically. On the other hand, qualitative data refers to data that sorts according to group, category, or type.

QUANTITATIVE DATA

- relating to numbers i.e. countable or measurable
- speaks to "how many", "how much", or "how often"
- data that is fixed and objective
- gathered by measuring and counting

QUALITATIVE DATA

- relating to language i.e. descriptive
- speaks to "why" or "how"
- data that is dynamic and subjective
- gathered by observing and interviewing

After you know what type of data you are working with, you have to form a connection between your data set. Is there a pattern or a developing relationship in the data? There are multiple ways of approaching this:

- Nominal - a grouping with no apparent order
- Ranking - an ordering of two or more groups in comparative significance
- Correlation - the positive, negative, or non-existing relationship between two or more variables
- Deviation - showing how the data relates to each other with an emphasis on the difference from the mean
- Distribution - how the data spreads out along a center
- Time-series - a series of data that is arranged around time, such as forecasting data

Use this relationship to help decide where a chart or graph is appropriate and where text provides a better explanation. Moreover, an effective strategy to minimize clutter but maximize information is to use a balanced mix of images and text. This balance makes your visualization clear and coherent.

Similarly, putting too much information or data in the same space drowns your viewer in visual noise. It has the same effect as trying to hear your friend while surrounded by people at a concert. You know they are talking to you, but the noise from the crowd plus the speakers drown out the message. Use the principles of design and layout to orient the components of your chart or graph to make it effortless for a viewer to navigate and comprehend. One trick to consider is to start with a gray or monotone design. Keep a bird's eye view of placement and flow first, then optimize with colors, patterns, and other elements. Focus on your essential elements, or the bones of your viz, and then add visual interest to force yourself to thoughtfully and purposefully regard all the components of your data visualization.

DATA TRANSPARENCY

One thing that makes an audience wary of trusting a data visualization is when there is no transparency around data sources. If you don't provide details on where your data came from, it is less likely that your audience will trust your work. Therefore, incorporate the habit of listing your data source in your visual to make your design credible and believable.

Listing sources is essential to gain trust and build a genuine connection with your audience. Without any trust, the opportunity to facilitate education and understanding can be lost.

KEY TAKEAWAYS

- Establishing a pre-flight checklist can come in handy to help you remember all the key points to cover in your design.
- A pre-flight checklist for a data viz may include defining your audience, clarifying your main message or information to share, and building a story around that information.
- Your audience should be your initial consideration when you plan your design.
- Providing context for your data visualization is essential for your audience to understand the story of the data and your perspective on it.
- Determine the type of data you have and any patterns or relationships in the data set.
- Many great data stories tend to follow three stages: introduction, key message, and conclusion.
- A data visualization must include your data source so your audience can build trust with you and your design.

CHAPTER 9

MAKE YOUR CHARTS PRETTY (AND SMART)

THIS CHAPTER FOCUSES on bringing together all of the information we have acquired up till now. Consider all you have learned about vision, perception, pre-attentive attributes, elements and principles of design, and layout principles. Incorporate them into your design while maintaining the narrative and aesthetics of the overall visual.

But are aesthetics that important if you have a functional design? I think so, yes. A viewer's general impression develops instantly before exploring further. People tend to gravitate to beautiful things, and pleasing aesthetics make a visual easier on the eye, attract a wide range of audiences, and give them a positive user experience. The aesthetic aspect of a design can encourage a particular mood, idea, emotion, or feeling in the audience. Furthermore, people are more patient, reliant, and forgiving when the design is aesthetically pleasing.

. . .

Creating a data visualization is a rewarding but laborious task that condenses information about lots and lots of data into one frame. By making it visually appealing and 'sticky' in the audience's minds, you can get them to stop, look, and explore.

HOW TO ENSURE YOUR DESIGN IS AESTHETICALLY PLEASING

In the words of the great architect and designer Mies Van Der Rohe, "Less is more." Think of your potential design as a blank canvas or screen. Every element added to the canvas becomes another element that your viewer's brain must process. Therefore, we must look deep into our narrative and focus on minimalism while designing for an audience. Ask yourself what you absolutely have to have in the visualization to convey your data story with enough context for the user. Remember that your viewer's minds have only a certain amount of mental processing capacity, and it is on you to design purposefully and smartly. Think of your finished data visualization as a future trip you have to pack for. You can only bring a small case with you to carry your essentials. What do you put in the case?

Think Like a Minimalist

But how do you choose which information to keep? Unfortunately, there is no straight answer because people make data visualizations for many situations and audiences. But an important principle we have covered that is crucial to minimizing cognitive load is to reduce clutter from your visuals. Clutter refers to those visual elements that take up space but don't aid in understanding. They can make a data visualization feel more complex than it is and drive the audience away from your design. Remember that our working memory can handle 4-5 chunks of information at a time before it has to work harder to understand. Creating a simple and minimal design will facilitate an unmistakable message to your audience and connect with them quickly.

Think back to the previous information on Gestalt's Principles. When distinguishing the essential elements from the clutter, consider closure, continuity, connection, proximity, and repetition while you design. Some other ways you can remove clutter and bring more attention to the visual is by:

- Removing unnecessary extra lines like grid lines, borderlines, and underlines
- Creating visual contrast by outlining shapes in a contrasting color, such as white
- Reducing the number of colors in the design or working to a limited color palette
- Using symbols or patterns instead of color
- Plan in black and white, then bring in color if you wish
- Utilizing negative space to highlight or contrast rather than thinking of it as empty space

Organize for Efficiency

The last few chapters have covered how to map out your ideas and pressed on how you can bring attention to focal points in your layout using tools such as the Gestalt principles and pre-attentive attributes. Pre-attentive attributes can signal the viewer what is most important and where they should look next. Using these attributes ensures that every element gets its time to shine without too much competition on the page.

. . .

Multiple methods can denote importance, but the most common approach is using color to highlight. Highlighting the component with a specific color or font makes that element stand out. Highlighting like this provides a clear signal of where the eye should go first, but this will only work if you have cut down the mess on the page. If you don't identify distractions and eliminate them, you will end up with a disorienting visual. Here are a few considerations that can help you identify these distractions so you can work on removing them from your design:

- Not all data is equally important. Create a road map of the main points for your story and only keep the required ones.
- Learn to summarize information when detail is not necessary.
- Look at each element with a critical eye and ask yourself whether it is essential. What role does it play in the visual?
- Keep less critical information in the background. Resist the temptation to stuff all of your work into one visual, especially if it serves no purpose.

Use Visual Hierarchy

Visual hierarchy is a means to signal to your audience how to read the page and in what order. You can do this using pre-attentive attributes (like we studied in earlier chapters). By utilizing these attributes of various strengths, you can lead the viewer through your visual the way you want to. Visual hierarchy attempts to take advantage of how our sensory memory works. Remember that sensory memory is automatic and unconscious. A viewer can scan a page and notice a break in a design without consciously looking for it. The human brain detects differences in patterns without making an effort or decision to look for them. Our eyes tend to look for the contrast of big vs. small, dark vs. light, and color vs. no color. Using these attributes to our advantage, we can create a more flowing narrative in our visualization.

Here are some extra attributes that are excellent tools for controlling a narrative: position, emphasis, and size

POSITION

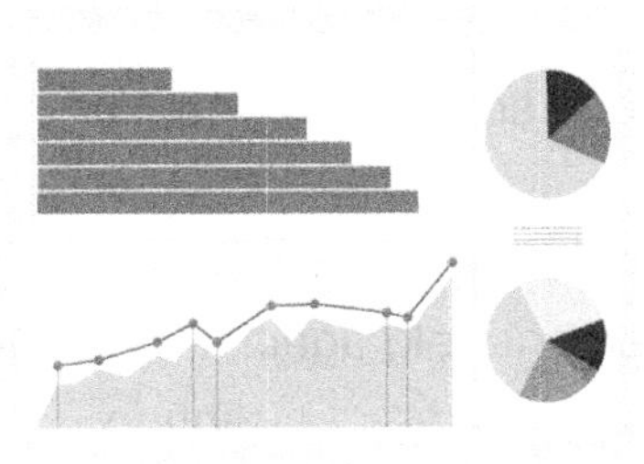

Position refers to the placement or orientation of images and text within the frame of your visualization. Thoughtfully placing these elements can guarantee visual hierarchy and guide the narrative's flow for your audience.

Most viewers will look at visualization from top to bottom, so choose what you want your viewer to see first and give it prime real estate in the visual.

Emphasis

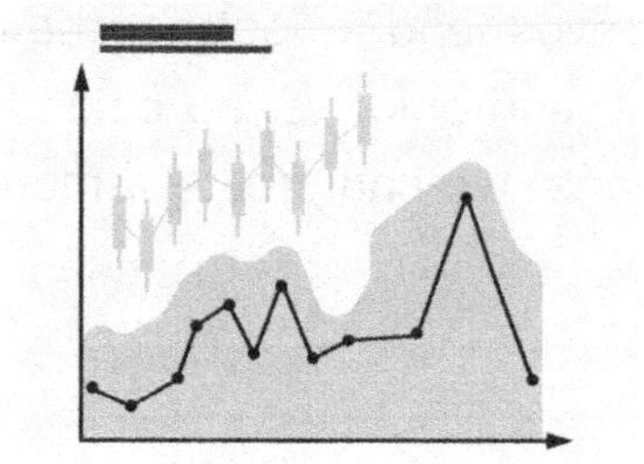

Emphasis is another way of establishing visual hierarchy in a visualization. You can achieve it in various ways, but it always involves utilizing contrast. Some of the ways to show emphasis may be:

- using white space around the visual or text to set it apart
- using a single color against no color
- making one element bold against other elements
- using varying sizes to direct attention

Size

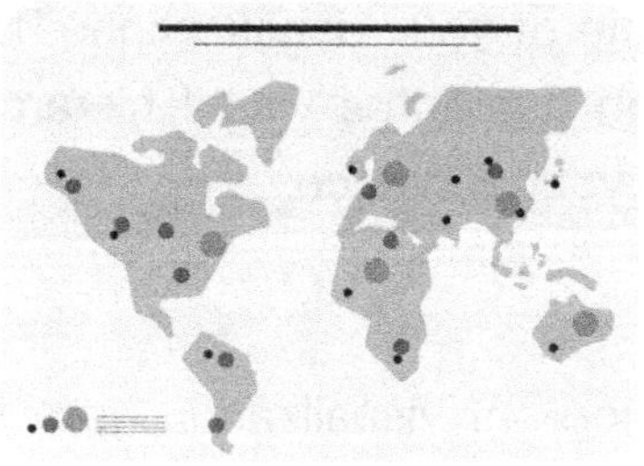

When you have multiple data points that are the same size, they can appear to blend into one big chunk, just as described in

Gestalt's similarity principle. You can rework this by using size to help visually differentiate data points and create emphasis that aids your viewer. If you don't want to use color, playing with varying sizes can be an alternative way to communicate to your audience. Contrasting sizing works because people have pre-determined assumptions of some elements in their minds. One example is with scaling values relating to the visual size of something. A smaller-sized circle may represent a small amount, while a big circle may represent a greater amount.

CONNECT WITH YOUR AUDIENCE

The sole reason for this book is to get you to start thinking differently and in the interest of your audience. Knowing and connecting to your audience is as valuable as being an expert in the elements and principles of design. According to Aaron Walter in his book '*Designing for Emotion*,' a visual or product has to be functional, reliable, and usable before making it beautiful. The end goal is to connect with the audience on a personal, emotional level and, if you can, invoke positive feelings in them. I keep mentioning positive emotions because they tend to ingrain positive memories in the long-term memory, which means that your visuals will be remembered and remembered fondly. Positive doesn't necessarily mean jumping for joy. It can also be a simple satisfaction or feeling of being able to relate.

PULL IT ALL TOGETHER

To sum it all up, if you want to design a data visualization that is functional, engaging, and aesthetic, use what I like to call the Grab-Guide-Ping method:

. . .

Grab

- Use purposeful pre-attentive attributes in your visualization to catch the audience's initial attention.
- Try using attributes such as size, color, contrast, and position since they trigger the sensory memory automatically.

Guide

- Group similar bundles of information into 4-5 chunks so that the cognitive strain is minimal and the brain can organize the information quickly.
- Use visual hierarchy to direct the viewer to the most important part of the visualization and then instruct where to look next.
- Focus on a minimal layout that creates a smooth and seamless experience that is easy to read and understand.
- Incorporate similar shapes to create uniformity and consistency within the design

Ping

- Ping the long-term memory by connecting emotionally with your viewer.
- Add anecdotes to share interesting data points.
- Allow the audience to put some of the pieces of your story together. Lead them to an aha moment or realization that makes them feel clever.

It may sound silly, but say it to yourself three times with emphasis on PING!, and I bet you'll remember it later. I understand that all these new tools and concepts may seem overwhelming and like a lot to think about at the same time. It is a learning curve, but with practice, you'll be able to incorporate these methods into your work and start seeing them in your peer's work. Learning how to create data visualizations that find a way into people's hearts (and minds) is the path to telling stories that people don't know yet but want to hear.

KEY TAKEAWAYS

- Aesthetic experiences draw people in. It is their first impression before they get deeper into your design.
- The aesthetics of a design can encourage a particular mood, idea, emotion, or feeling in the audience.
- Think like a minimalist and try to reduce the number of elements in your visualization to the most essential.
- Zoom out during the design phase and consider the whole composition. Use attributes such as position, emphasis, and size to create a visual hierarchy.
- Use design to GRAB your audience's attention, GUIDE their eye through your visual, and PING their long-term memory by creating a positive experience.
- Positive emotions tend to create longer-lasting memories that connect with the viewer and save in the long-term memory.

CHAPTER 10
ACCESSIBILITY, DIVERSITY, AND INCLUSION

IN THE CURRENT WORLD, we are way more connected than our predecessors were and continue to interact with each other through technology almost every day. More focus than ever is on making products, workplaces, and experiences accessible and inclusive. But what does it mean to incorporate accessibility, diversity, and inclusion in your design?

Accessibility and inclusion in design refer to the need to consider all potential viewers. When we design for our audience, we often tend to forget that our visuals might be reaching people who are colorblind or have visual, auditory, motor, and cognitive disabilities. While trying to wrap up a project, you can forget that you are presenting to a much larger and more diverse audience than you may think. It is a best practice to make accessibility and inclusion a part of the process while designing a visual.

QUESTIONS TO ASK YOURSELF BEFORE STARTING THE DESIGN PHASE

This book focuses primarily on the design of a data visualization, but this chapter focuses on your audience's experience. Their experience starts the minute you begin planning your visualization. Designing for accessibility can sometimes be a couple of checkboxes that you might tick off to pass your visual as inclusive. But the real problem is that accessibility is less about compliance and more about usability.

One way to ensure that you remember to include accessibility in your design is by answering these who, what, where, why, how (W5H) questions before you start designing:

- Who is viewing your visual?
- What are they viewing?
- Where are they viewing it?
- When are they viewing it?
- Why are they viewing it?
- How are they viewing it?

As we mentioned in Chapter 9, a good design invokes a positive or connecting feeling in the audience. And when you design with accessibility in mind, even viewers with specific disabilities will look at your visual, perceive it and store it in their long-term memory. You might be thinking, "is it even possible to design a visual that creates an experience for every kind of audience?" Well, it might not be. But you can aim to create experiences through your design that everybody can access. The key thing to remember is to design with sensitivity and empathy.

. . .

Thinking of accessibility and inclusivity can be challenging, especially if you are unsure where to start. One activity you can undertake is drawing out an empathy map before beginning a visualization. Empathy maps are widely used in UX/UI to help designers understand their audience's needs. They force you to think about what your viewers might say, think, do, or feel when they see your visual. It is as simple as drawing a circle on paper to represent your viewer. Draw branches out in four directions from the circle to divide the page into four quadrants. These four blocks are spaces to assign

- Says - how your viewer speaks about your visual to themselves and others
- Thinks - what your viewer thinks to themselves when they look at your visual
- Does - action your viewer will take after looking at your visual
- Feels - how your visual will make your viewer feel

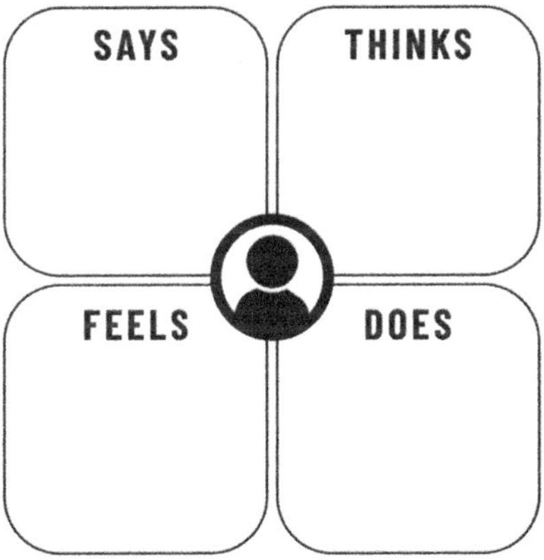

When you start your project, you can infer your viewer's responses. However, as you work on your visual and get feedback from your peers, record what they say, think, do, and feel in your empathy map to see if there are differences between where you want to be and where you are.

HOW TO CATER TO A VISUALLY CHALLENGED AUDIENCE

If you haven't thought about accessibility in your design before today, it's alright. There is time to change, and we all do it at our own pace. When you think about accessibility, one thing to remember is that people who have visual, motor, speech, auditory, or cognitive disabilities tend to rely on other assistive technologies and interaction methods with people in their everyday lives. Therefore, if you want your design or visualization to be inclusive, a few ways you can do that are by:

- Naming image files with detailed descriptions for people who are visually impaired and use narrative devices or assistive technologies to navigate social media or the web.
- Running your visual through a color blindness simulator online. The most common color blindnesses are red-green and blue-yellow.
- Using symbols, patterns, or shapes instead of colors, making it easier for people with certain disabilities to differentiate values.

We might not realize how these minor changes and considerations in our design can create a drastic difference for our audience. For example, there was a great outrage over the 2016 NFL season when the two competing teams had red and green jerseys on. Fans worldwide thought this was a massive problem because the viewers who were either fully or partially red-green colorblind could not differentiate between the two team's players.

. . .

While red-green is the most common, let's also speak to the other less common color blindness/weaknesses you should remember when you design.

Red-Green Color Blindness

This color-blindness is the most common globally and occurs due to the loss of red cone and green cone photopigments. There are other subtypes within the red-green color-blindness:

- **Protanomaly:** In this disorder, the red cone photopigment is abnormal, and all red-yellow tones appear green.
- **Protanopia:** In this disorder, the red photopigment is absent, and all tones of red appear black, while orange, yellow and green appear yellow.
- **Deuteranomaly:** In this condition, green cone pigment is abnormal, and it becomes extremely difficult to differentiate between purple and blue. Green and yellow also appear red in this condition.
- **Deuteranopia:** There is a loss of working green photo-pigments in this condition. The reds appear brown-yellow, and the greens appear beige.

Blue-Yellow Color Blindness

This type of color blindness is rarer than the red-green type. In this condition, the blue photopigment is either missing or abnormal. There are also two types of blue-yellow color-blindness:

- **Tritanomaly** is an extremely rare condition in which the blue cone cells have limited function. In this disorder, blue appears green and yellow tones appear red or pink.
- **Tritanopia** is an autosomal recessive disorder and occurs in females and males equally. It is extremely rare, but blue cone cells are absent in this condition. Blue appears green, and yellow looks like violet or gray.

Complete Color Blindness

People with complete color blindness don't experience colors at all. Their total vision is monochrome, which often also affects the clarity of their sight. People with partial or complete color blindness have extreme difficulty differentiating between colors. Most of the colors look the same unless there is a difference in the contrast of the colors.

It can be tricky, but it is your responsibility to consider your audience and present your idea, story, or information to them in the best possible way. It has been made extremely easy to be inclusive in this day and age. For example, multiple websites, browser extensions, and apps with color blindness simulators are available to scan your visual and determine whether it is suitable for people who are colorblind or have other disabilities. Another approach to making it easy is to view your visualization in monochrome. If you can still distinguish the different values in black and white, you are good to go.

. . .

Accessibility and inclusivity are also about the words and phrases used to describe differently-abled people. Being sensitive to the words you use and only using phrases recognized and accepted by your audience is also a considerable part of accessibility in design. Consider using "people-first" language, for example, to refer to a "person with disabilities" rather than a "disabled person." The whole point of this discussion is that a design is empathetic if developed with care, sensitivity, and respect.

KEY TAKEAWAYS

- Accessibility is less about compliance and more about usability.
- W5H is a great exercise to undertake before you start the design process.
- Empathy maps are a tool that can help you in knowing and understanding your audience better.
- People with visual, motor, speech, auditory, or cognitive disabilities rely on assistive technologies that you can accommodate in your designs.
- Red-green color blindness is the most common.
- Yellow-blue color-blindness is rarer, but also a color combination to be wary of using if possible.
- People with complete color blindness don't experience colors at all.
- Inclusivity is also about phrases and word choice. Use terms accepted by your audience and consider "people-first" languaging.

CHAPTER 11
A STYLE GUIDE – THE DESIGNER'S SECRET PLAYBOOK

NEXT, let us move on to how you, a budding designer, can create a repeatable framework called a style guide, so you don't have to build your visualizations from scratch every time you approach a new data viz project. But before we teach you how to make this guide, let's quickly jump into what a style guide is and how it can benefit you.

WHAT IS A STYLE GUIDE?

Marketing teams for brands often use style guides to provide employees with a set of rules to follow when creating content or promotional materials. These guides are living documents used to maintain consistency and reduce decision-making. A data visualization style guide works in the same way. It provides a base set of best practices and style decisions and breaks down a visualization into components with instructions for how and when to use them in your visual. The purpose of a document like this is to house all of your repeatable design choices, so you don't have to remember something like what font you used or what color palette you prefer every time you start a new assignment. Instead, you have more time to focus on the more project-specific aspects of your viz.

Style guides are beneficial for every organization that produces data visualizations. They help provide a detailed guide for team members on what they should and should not include in their visualizations. New team members use these guides to familiarize themselves with the company's visual direction and branding values. In addition, it sets the tone and expectation for the others in the organization that data visualization is just as valuable to the business as other branding materials.

A style guide helps make a data visualization stand out whether you are an individual freelancer or working for a large organization. It helps you work in a specific style that keeps your choices consistent and efficient while building your brand. Consider creating a style guide for yourself. You can even create multiple guides for different project types. Perhaps you want a consistent look and feel for your personal work and another set of instructions for client work? It will significantly decrease time spent on design, reduce decision fatigue, and make your work more identifiable to your audience by building your unique brand.

THE PARTS OF A DATA VIZ STYLE GUIDE

Working to a style guide means sticking to a specified set of choices for layout and the elements within the composition. Take whatever you have learned until now about elements and principles of design, principles of layout and arrangement, preattentive attributes, and Gestalt principles and apply them to your style guide. Choose things such as a color palette, graph types, font set, and basic layout rules that you resonate with, and use them as a foundation to design your visualization.

A good, comprehensive data visualization style guide breaks down everything on a graph, chart, and table and explains the best practices and design strategies to incorporate into your visual. Here are some of the features your style guide could include:

Graph Anatomy

Graph anatomy refers to the placement of labels, titles, and other elements in a composition. It also focuses on the overall dimensions of the graph or chart, including the output size. Your output size will differ depending on the type of graph or chart you design and whether you develop it for print or web. Other elements that come under graph anatomy are

- figure number
- dimensions
- title and subtitle
- axis labels
- gridlines
- sources
- notes
- logo
- legends

Color Palette

This book has broadly studied color and its importance in data visualization. Color has undeniable power in a visualization and is probably one of the top things that stand out to an audience. Every successful brand has a brand color palette that they use in their visuals to distinguish them. Can you imagine Coca-Cola without the red on their cans?

The correct use of color can make or break a visualization. It is highly efficient in highlighting a point in your narrative but must be used thoughtfully. Since color can be overwhelming with all the choices, I recommend starting with three to build your first palette. These three colors will have specific roles:

1. Main or Key Color
2. Secondary Color
3. Accent Color

Your main or key color will be used primarily throughout your visualization, with support from your secondary color. It is up to you to decide how similar in hue you want your first two colors, but make sure your accent color contrasts the first two. Use an accent color to make things pop out or appear highlighted to create emphasis. Employ it sparingly like an exclamation mark.

An example of this three-tone color palette could be black, white, and red. It may seem incredibly basic and limiting, but remember that you can use tone and opacity to stretch how you use these three colors. I am sure you think red is the only choice for an accent color, but imagine a sweeping red background with black text. A white shape popped against the red background or a white outline around the black text is also a great accent. Keep your design mind open!

Typography

A data visualization style guide must include a list of all the usable font styles and their size. No more than two fonts are necessary for a visualization. Sans-serif fonts are generally the best choice since they are easier to view on devices or the web. Avoid handwritten style fonts when possible unless it is appropriate for the data. However, in a business setting, it is best to stick to simple and easy-to-read fonts.

Like with color, it is better to set limitations with fonts and sizing to keep your visual accessible to read. You want to make sure your audience can read your text, whether displayed in a dark room via projector or on a small phone screen. Using the same font with different styles can make the design look cohesive and coherent. Utilize your typography's sizing, boldness, and proximity to create hierarchy and emphasis in your visual.

Graph Types

Another essential thing to add to your data visualization style guide is a set of instructions or details about particular graphs or chart types. Your organization or personal design philosophy may prefer specific charts or graph types as a go-to. In contrast, others may be omitted or denied.

. . .

Exporting Images

When your design or visualization is ready to be viewed by your audience, you must export your work correctly. Choosing the suitable file format is crucial, and deciding between a raster image and a vector image must be made according to your design and output space. Is this visual meant for print? If so, export your visual based on the needs of your printer or printing company. If you are exporting for the web or sending via email, you'll want to make sure your file size is small enough, and you correctly compress your images. The best strategy is distinguishing different approaches for different scenarios and determining what works best for your needs.

Accessibility, Inclusion, and Diversity

Previously, we talked about the importance of accessibility, inclusivity, and diversity in a design. Even though there are few specific rules about how to include accessibility in design, it is in your best interest to highlight it as something to build into each project. One example could be to write detailed image labels as the file name of your images. When a person using a device assistant comes across your design, they'll hear your description rather than just Image.jpg. Another practice could be to run your visual through a color blindness simulator, as mentioned in the last chapter.

What everyone can do is learn to design with empathy and consider how people with disabilities access your content. It is imperative to take the time out to evaluate what approaches can make this world a better, more inclusive space for everyone.

There is no right or wrong way for you to create a style guide. It is a personal, living document that can be updated and modified according to your needs. A detailed style guide is made to keep you organized and must benefit your workflow rather than take away from it. It helps to set the tone for your entire data visualization and, on a bigger scale, characterizes your brand and identity to your audience.

KEY TAKEAWAYS

- A data viz style guide helps provide detailed instructions for team members or individuals and educates them on what elements a data visualization should or should not include.
- The purpose of a document like this is to house all of your repeatable design choices, so you don't have to remember them every time you begin a new project.
- Working with a style guide helps maintain visual consistency and speeds up the design process.
- Some of the basic components a style guide should include for data visualization are details on graph anatomy, color palette, typography, graph or chart types, exporting, and accessibility.

CHAPTER 12
DESIGN PROCESS

A SINGLE-PAGE LAYOUT that can be looked at, read, and digested within a minute of viewing can take much longer than people think to design. The complete design experience is a multi-phase process before you finally end up with a finished visual and put it out there for your audience. Assembling the parts of the layout in a visualization is a continuous process that spans from the concepting, drafting, to final fiddling.

WHAT IS A DESIGN PROCESS, AND WHY IS IT IMPORTANT?

A design process refers to steps made to produce a meaningful visual. It is very similar to a style guide in that it has a particular order that you should follow to maintain consistency. A design process is not the same for every creator. For example, how someone cooks a meal will differ from person to person. Maybe I start by gathering all my ingredients together before I start. Meanwhile, you might grab each ingredient as you need it. The result will, of course, be the same, but the process might differ based on preference and experience.

. . .

Adhering to a repeatable process helps improve your work, saves you time, makes the workflow smoother, and helps maintain consistency in your output. Moreover, if you work in a company with a team, having a design process helps eliminate any decision-making that might waste time. And if you follow a repeated, trusted process every time you design, it increases efficiency and enables you to finish your projects faster.

A DESIGN PROCESS STEP BY STEP

Some of the basic steps that a workflow follows during the design phase are:

The Project Brief

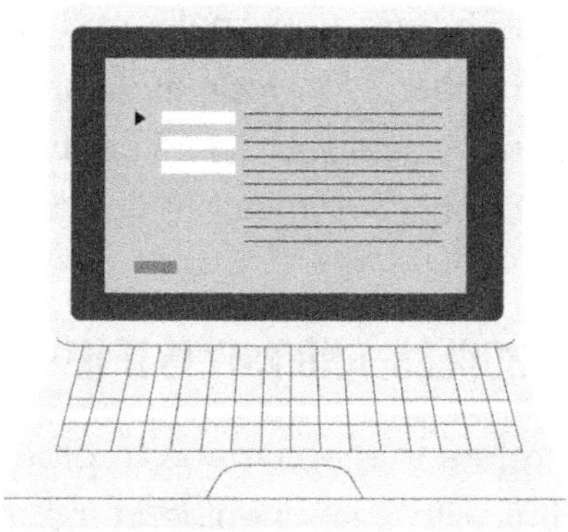

The project brief is just as it sounds. It is a short description of the project and the requirements of the final deliverables. In the brief, you will want to note all those things we mentioned in the pre-flight checklist, such as who your audience is, the purpose of this visualization, and what story you want to tell. This brief can be provided to you by your boss or client, an online community data challenge, or you can make one for yourself.

. . .

The project brief is where you can let your imagination run wild and think of everything you can do to make your visualization exciting and engaging. You will need to determine what style guide you want to follow, if any at all, according to your concept and data. It is alright to stray from the brief for a personal project if you find your data leading you in a different direction. However, if the project is for a boss or client, you need to take the brief seriously and use it as a blueprint for your design.

Preplanning - Storyboarding

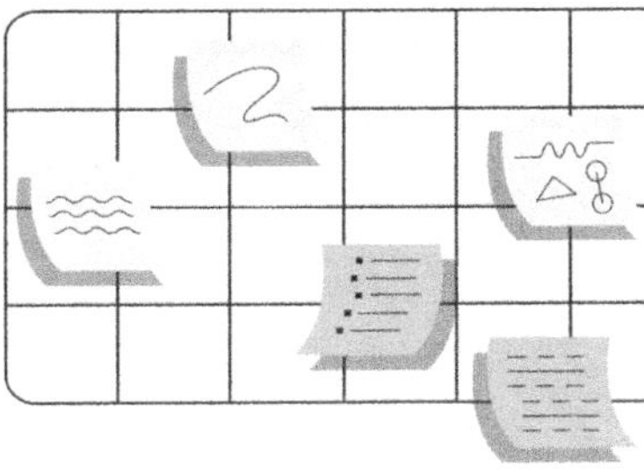

After reading the project brief, you can move on to the actual design process. The brief must have inspired some ideas and concepts, which you now put to paper in the form of sketches. No matter how solid you might think your idea is, it would be best if you sketched it out. The mind is extremely good at glossing over faults and logical loopholes in your design that you can catch early and fix.

Start with the basic shapes or chart types that could work for your data. Putting your thoughts on paper can show if you need to make adjustments or different. These drawings are generally just brainstorming and do not have to look perfect and professional. You must remember that because your design works in your head, it may not work in reality. Taking the time to sketch out initial ideas will help you identify any issues right away and save rework down the line. Another method to follow is imagining the potential visual and creating a storyboard to create structure and follow the flow of the narrative you want to tell. Keep this step low-tech. Try using sticky notes or small pieces of paper to lay out the parts of your narrative. Having this bird's eye view will give you an overall perspective and help you see if there are any gaps in the story. It becomes a best practice to do some quick study sketches of your idea to have a solid base from which to work. This trick can provide an objective perspective to your design choices and keep you from becoming too emotionally attached to an idea that might not be best for your data or audience.

Drafting

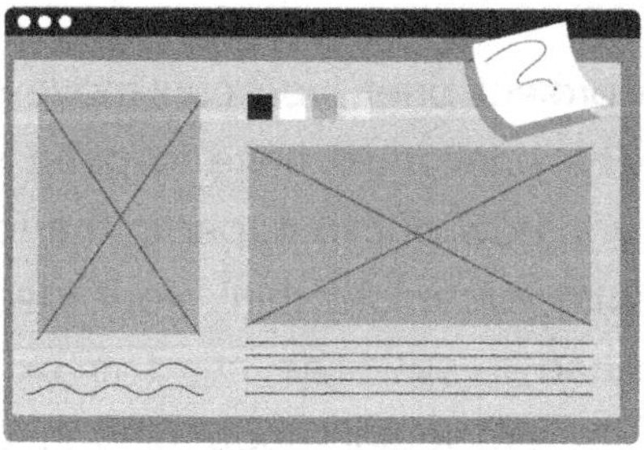

Your favorite ideas at the start of the project don't always play out in the end, but that is okay. Because these ideas are a starting point that leads you to your final design, they are still important. In the drafting stage of the design process, you have settled on an idea or sketch. Now you work towards developing the layout and composition of your visualization with your chosen tools. This is the most fun part of the design process because you have the creative freedom to play around with the layout and how you arrange the elements within. Experiment with combinations of attributes to see how you can best represent your data.

Final Output

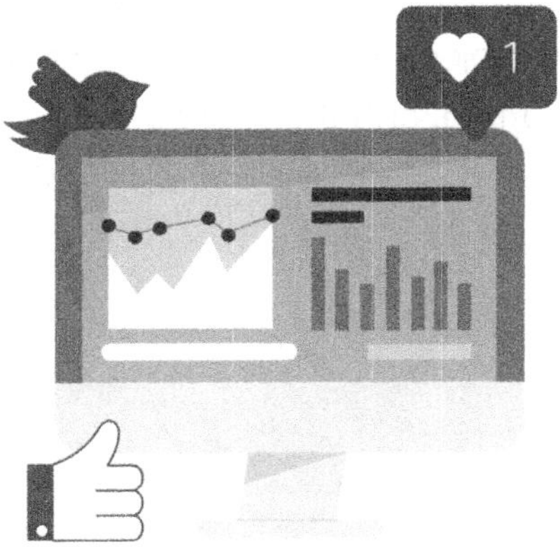

After you have successfully chosen a layout and designed your visualization, it is time for you to step back...almost. One crucial step of the design process, as we learned in Chapter 11, is to export your image/visual in the required format, whether it is a presentation or a poster. Now you can share your final output with your client or your target audience and celebrate all the hard work you have put into it. Share your work with the community online for feedback if you choose or add it to your portfolio. But don't forget to celebrate. You finished it!

KEY TAKEAWAYS

- A design process refers to the series of steps taken to produce a meaningful visual.
- Adhering to a repeatable process helps improve your work, saves you time, makes the workflow smoother, and helps maintain consistency.
- If you follow a repeated, trusted process every time you design, it increases efficiency and helps you finish your projects faster.
- The basic steps you follow during the design phase are acquiring the project brief, concept planning or sketching, drafting, and the final output.

PART FOUR
MOVING FORWARD

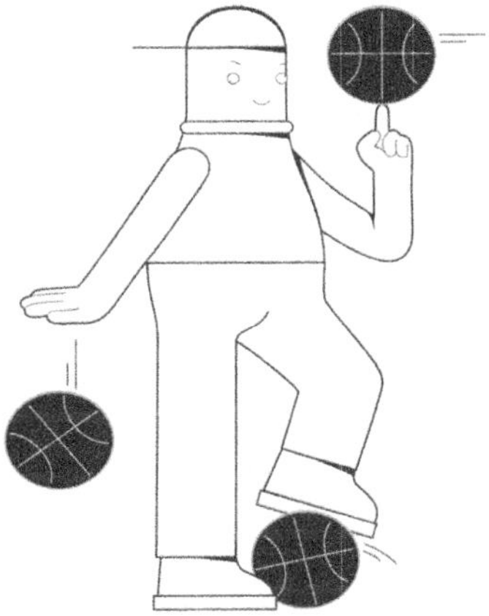

"Talent is a pursued interest. In other words, anything you are willing to practice, you can do."

BOB ROSS

CHAPTER 13
STEAL LIKE A RACCOON

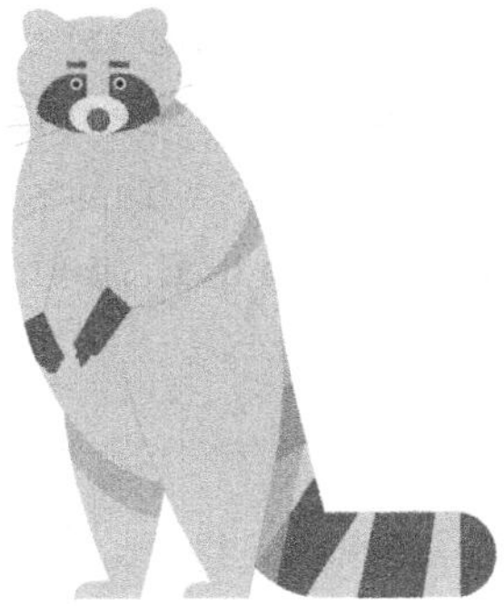

WHEN YOU LOOK CLOSELY at a raccoon, do you notice the bandit eyes nature has given them? The black mask makes them look mischievous but adorable at the same time. You can find these creatures foraging in your backyard garbage for food, but also shiny, glossy things. These shiny things are curious for raccoons. They take them home to add to their collection. I believe this habit of collecting things that delight and intrigue is a shared characteristic between raccoons and humans.

INSPIRATION IS EVERYWHERE

I say this because I believe when a person finds something that captivates them or inspires them, they tend to use it as a muse for their work. So, for example, when a musician gets inspiration from an artwork, a film, or a book, perhaps, they use it to create music and further inspire someone in their line of work. Similarly, when designers look at other posters, visualizations, or magazines in print or online, they collect those nuggets of inspiration and use them as references later.

In the world of design, gathering inspirational content is very important. Approaching the design process can be extremely daunting and intimidating, especially for beginners. Therefore, you need to look at what others have done so you can start to create something that makes you feel proud. Ernest Hemingway read Shakespeare and Dostoyevsky, while Picasso studied the works of Velazquez and Goya. Take inspiration from the process of these past masters. While we might not be Picasso, the core principle is the same. To be good at something, you have to understand the basics and draw inspiration from past pieces of visualization that you have seen and enjoyed.

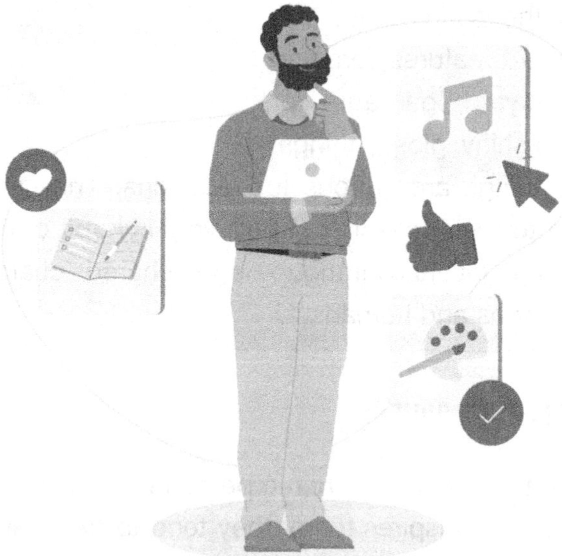

Gathering inspiration content from the past does not mean your visualization will be a copy or imitation. There is no need for us to reinvent the wheel. People tend to respond better to things they already know. So, when you come across a data visualization that you enjoy or admire, save it in your archive or a folder with a note about why you enjoyed it and what you want to take from it. Keeping notes about what you like or don't like from your inspiration will help you learn what you like to visualize and, more importantly, how.

All creative work you see today or will see tomorrow is built on what came before it. There is nothing 'original' in this world, and people who say their work is "100% original" are unaware of the sources or just plain fibbing.

> "*What is originality? Undetected plagiarism.*"
>
> - WILLIAM RALPH INGE

KEEP A "SWIPE FILE"

Collecting inspiration not only enables you to learn from different artists' work but also helps you out on a rainy day. Gather inspiration from wherever you like and put them in what artist and author Austin Kleon refers to as a "swipe file" – a file you keep filled with the stuff you have swiped for inspiration. This swipe file can be digital or analog and reanimate those ideas, principles, and designs in your work.

Why is it essential to keep a swipe file, you ask? When you look at the work of other artists, you look at it critically and break down the elements in the visualization. This practice gains you a fresh perspective, and you learn how to do certain things that you may not have tried in your work before. Our brains are constantly learning, observing, and analyzing data. The more you look at something and analyze it, the more your brain realizes what is noteworthy and why. It starts sorting through the details and notices things that stick out, storing this information for future use. Keeping a collection of inspirational content ensures that your brain builds new connections. It encodes all the lessons you are subconsciously or consciously learning to store them for future use.

KEY TAKEAWAYS:

- In the world of design, gathering inspirational content is a fundamental practice.
- All creative work you see today or will see tomorrow is inspired by what came before it.
- Keeping a collection of inspiration will enable you to learn from different artists' work and help you out when you need some fresh ideas.
- The more you look at a visual and analyze it, the more your brain comprehends the layout and visual composition, after which it stores this knowledge for future use.

CHAPTER 14
REMIX LIKE A DJ

IN THE LAST CHAPTER, you learned how to "steal" inspiration to save and look at later, but in this chapter, we will focus on effectively deconstructing inspirational content so that you can understand why it is successful. Now you take all your gathered material and learn from it. By keeping this material together, you can start seeing repeating themes or elements that excite you. Generally, just a piece of another work catches your eye rather than the whole thing. What pieces from your swipe file content do you find inspiring?

To become an excellent data storyteller, you must look at other data stories. Only when you let yourself be an enthusiast of the craft can you truly succeed in making informed and beautiful data visualizations. A visual created for a research presentation will look very different from a graphic designed for journalism. When you deconstruct other visualizations, you get to pick and choose the elements you want to keep for yourself. Try reproducing those things you like in your own work to practice and understand the decision-making behind the visualization.

Walking in the shoes of another designer will enable you to see the world differently, and that is how you will grow better and more expressive in your designs.

PICK IT APART

The first thing you want to do is look at all the images, visualizations, illustrations, or designs you may have collected for your inspiration archive. Look closely, and analyze each one. What made you choose this against so many other visuals? Why does it inspire you? And what aspects of this beautiful work do you like? There can be many things to take away from a visual. You might like the color palette chosen by the designer or the excellent use of space, annotation, or font. You may be interested in how it cleverly uses contrast or visual hierarchy. Breaking down the visual into smaller parts may enable you to understand how it is put together and incorporate what is successful in the design into your own projects.

In a visual, you may find multiple elements that inspire you or layout solutions that grab you. When you find an inspirational image, note all the parts you like and figure out why they work so well. There is no need for you to feel daunted. Keep it simple by focusing on just one or two elements at a time, such as graph or chart choice, use of negative space, or label placement. Transform these elements into new project ideas. Apply them to a previous project or even attempt to remake a visual you are inspired by. How could you improve it? Visual ideas are resilient and expansive and insist on us to break them down and learn from them so you can create a better, more comprehensive, and inspiring piece. This questioning and reworking of content you are inspired by will lead to your own aha moments and the seed of new growth in your personal development.

Stay playful as you rework the elements from your inspiration into your experiments. Get creative by combining it in different ways for the same project. This offers an opportunity for us to consider various ideas and new content that we might be able to create from the inspiration. For example, you can rework the same data set with different graph types or create multiple layout options for the same set of components. Imagine a picture of a loved one displayed as a photograph, painting, and finally, a black and white drawing. It is the same image shown in three different ways, but can you see how each makes you feel different? Would you want to hang the painting or drawing? Does one version feel more special? Small changes can have a big impact. Try to see how creating different versions of your visual by adjusting one element at a time can produce that effect.

EMULATE, DON'T PLAGIARIZE

When you look for inspiration in other people's work, beware of copying or blatant plagiarism. You are free to borrow or look at an idea from a visualization to develop it for yourself or find a way to use it in your visual. There are multiple ways to interpret it in your design, making it your own. Try to adapt the elements to your style differently from the source inspiration to get the most practice. This process is known as practice, not plagiarism. Plagiarism comes in when you take someone else's work and ideas and try to pass them off as your own. Taking inspiration works like a mechanic taking apart a car to understand how it runs and learning by reverse-engineering. The point of working from inspiration is not to see and repeat but to analyze and understand.

> "Those who do not want to imitate anything produce nothing"
>
> - SALVADOR DALI

Learning from the masters is the most promising way to learn, but if you worry about copying and want your work to be more authentic, consider setting up boundaries for yourself to work within. When you work with an element and try to put it into different designs, you can spend hours, days, and even weeks coming up with different results. How do you know when to stop? If you work with a boundary or set a limitation, you restrict yourself from going too far. Endless possibilities may sound ideal but can actually cause analysis paralysis. These limitations may be a good challenge for you by sparking creativity by making your brain go into problem-solving mode. Some examples of constraints to incite creativity are:

- A deadline or a time limit within which you have to finish a visualization
- Working to a style guide set up by you or your client
- Designing for a data challenge
- Keeping it minimal and utilitarian
- Making it as expressive and artistic as you can
- Only using color
- Only playing with a monochrome palette
- Choosing to highlight a specific element or design principle in your visual like line, shape, balance, or contrast

KEY TAKEAWAYS

- Analyze the chosen inspirational work and break it down into details around what you appreciate and why.
- To become an excellent data storyteller, you must look at multiple data stories across disciplines, from research to data journalism. Observe and take in other people's work.
- Visual ideas are resilient and expansive. They insist on us to break them down and learn from them so you can create a better, more comprehensive visual.
- When you look for inspiration in other people's work, beware of copying or blatant plagiarism. Plagiarism comes in when you take someone else's work and ideas and try to pass them off as your own.
- The point of working from inspiration is not to see and repeat but to analyze and understand.
- Set boundaries or limitations as you work to avoid analysis paralysis and encourage your brain to go into problem-solving mode.

CHAPTER 15
PRACTICE LIKE AN ATHLETE

IN THE PREVIOUS CHAPTER, we focused on the importance of collecting inspirational data for your growth and practice. Whatever you find appealing in one visual can be learned from and replicated in your work. To continuously improve your skill set, you must develop this practice of observing, learning, and applying to maintain that conscious design mind.

This is important because, in the field of design, moments of inspiration come in bouts. Sometimes you might feel demotivated or uninspired to think about design and let that extinguish your creative drive. Other times a strong dose of spontaneous inspiration washes over you, and you feel invigorated and possessed to work. But these moments are few and often short-lived. Motivation brings you to the starting line, but discipline and consistency take you through the finish.

. . .

A habit like brushing your teeth or making a coffee takes less energy to perform compared to a task you only do once in a while. Remember that our long-term memory stores knowledge that has been repeated and practiced, making it easier to recall. Repeating fundamentals over and over builds muscle memory and creates a core foundation of knowledge. The field of design or data visualization may require you to be creative with your practice. There is no one-size-fits-all approach, but instead, start with knowing yourself and how you like to work. Go through the basics, and learn from all the visuals that inspire you. From here, you can deconstruct what you have collected and rebuild it back like a puzzle based on your expanding design philosophy.

> *Inspiration is for amateurs. The rest of us just show up and get to work.*
>
> - CHUCK CLOSE

HOW TO ESTABLISH A HEALTHY PRACTICE ROUTINE

How does one move forward when there is no motivation or inspiration to get you started? The first step would be to show up and make an effort. It would be ridiculous for you to think a professional athlete enters into a championship if they only practiced or trained when they felt like it. The profession requires them to get up, show up, and make an effort to have a chance of winning. Design work is not so different. If you only practice your skill when inspiration hits or when you are bored and have some time on your hands, you will practice randomly, and ultimately, it will result in the slow death of your creativity.

. . .

Consistency is Key

Consistency is critical to improving your designs. A creative who works to a consistent and frequent practice session gets progressively better over time, and ultimately it shows in the way they design. If you have trouble motivating yourself, a strict work/practice schedule can be highly productive in the long term. If you keep your practice schedule too flexible, you will find ways to avoid it. There will always be something to check on your phone or an email that needs a reply. Instead, having a scheduled block of work time will force you to remain motivated and show up. Starting with something as simple as showing up will keep you practicing despite any mental block you may encounter.

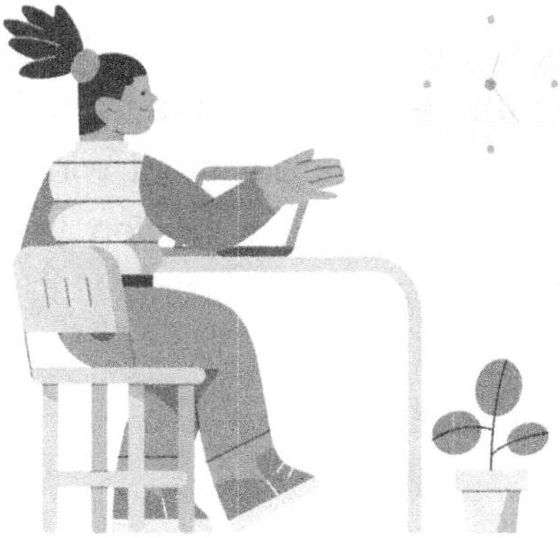

Familiar work setups and spaces get you in the zone faster.

There are some tried-and-tested methods to keep you consistent. For instance, having or creating a comfortable desk space for yourself can play a massive role in establishing a practice habit. Working in a familiar environment or setup relaxes your brain, and therefore you will not waste time settling into your surroundings. Instead, you can focus solely on your tasks until adequate time for your work block has passed. As you learn, you make fewer mistakes and spend less time reworking. You design faster and faster as your brain creates new connections and skips over unwanted and time-consuming trial-and-error decision-making.

In this age of hustle, we fall under the pressure of productivity and fear of underachieving. When we are not motivated, we tend to look at other people working and thriving, and instead of taking them as inspiration, we compare them to ourselves. Comparison can be a harmful thing. Every individual has a particular technique, style, or method that they work with, and comparing yourself to another will only diminish your efforts. Incremental changes and progress over time will yield major results. Remember that everyone starts from the beginning and those who practice and stay consistent win in the end. When you are constantly practicing and honing your skills, you will ultimately see improvement in your work.

Avoid Overwhelm

Another thing that people often experience is that continuous practice of your craft saves you from the prospect of "creative flooding." Creative flooding is the opposite of a creative block. It is the increase in the creative flow beyond what is manageable. When creative flooding happens, you may feel overwhelmed with ideas, and as a result, your flow reduces or stops as you try to cope with not knowing where to start. Too much choice can be overwhelming and too much to handle. During these times, you might want to establish some creative 'control.' Creative control implies prioritizing, organizing, and managing all that one has to do or practice. It often involves establishing rules and ways to manage time. Ultimately, you are looking for a creative discipline that only you can set to prevent burnout. Some things that you may want to try to cultivate a disciplined practice are:

- Time management
- Working with a design process (as explained in Chapter 12)
- Having a consistent routine
- Setting deadlines within the project
- Focusing on one task at a time
- Looking for an accountability partner

These are only considerations. Every person has a different way of processing, practicing, and working. The key is figuring out what works best for you. Try various things and keep what works for you. Eliminate what doesn't until you reach a routine that you favor. It is okay for you to feel frustrated in the beginning. Keep honing the practice until it becomes a habit.

Creating a data visualization, much like all forms of design, can be an individual and personal process. Each data author has their own unique way of visually organizing their data, which can be both intimidating and exciting. It is a great idea to get constructive criticism from friends and peers to keep learning and growing.

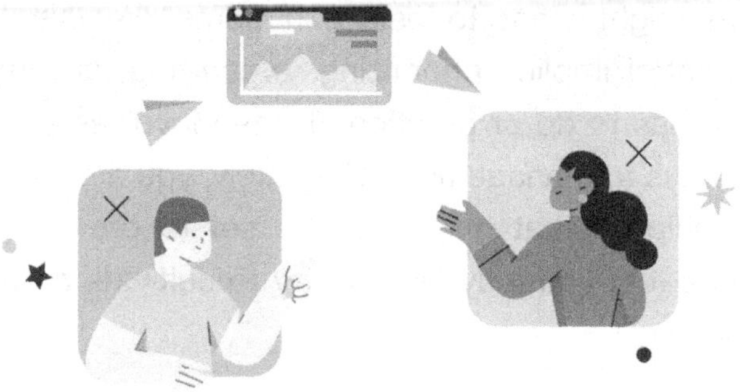

Feedback from peers can be an invaluable resource for growth.

Consider forming a community with your close friends or colleagues or joining a cohort to learn together and bounce ideas off each other. An extra set of eyes to provide feedback will quickly improve your work. Community fosters growth, and the more you learn from your peers, the more you practice, and the better you get!

KEY TAKEAWAYS

- Good design demands study, attention, and constant practice from us.
- Practice turns your understanding and love for something into an experience and builds on that experience to create a foundation of knowledge.
- Consistency is the secret to success. Consistency creates discipline and habit, which in time, yield results.
- Creating a comfortable desk or workspace for yourself can play a huge role in making it easy and enjoyable to get started with your routine of practice.
- Creative flooding is the increase of the creative flow of ideas beyond what could be manageable and leads to overwhelm. A lack of routine can make you swing from creative block to creative flood. Avoid this by playing with different ways to build a repeatable process that you can maintain.
- Community fosters growth. Share your work in person or online and ask for feedback so you can improve and grow with each project.

CONCLUSION

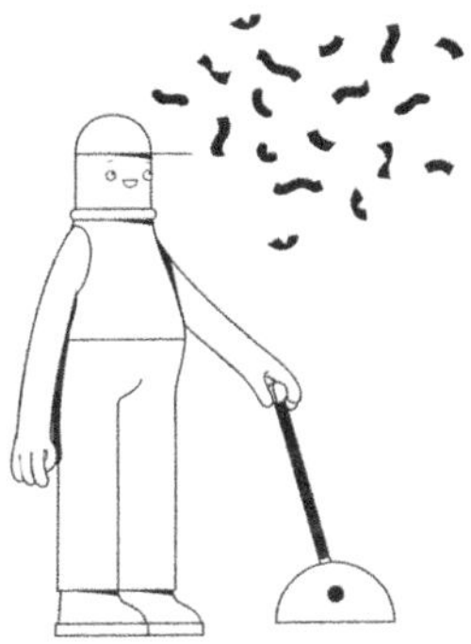

Congratulations on having made it so far along in the book. By now, you must be pretty well acquainted with the entire process involved in creating a successfully designed visualization. But to recap, the design process is more than what meets the eye. It is what greets the eye, is processed and understood by the brain, and hopefully connects emotionally to be remembered for a long time after. It is a process that truly begins and ends with human perception.

As we studied in the first few chapters, everything about your design has to cater to who your audience is and how they will be experiencing your visual. Whether using pre-attentive attributes to make the composition more eye-catching for an important presentation, educational and straightforward for a research document or fun and experimental for a data viz challenge, a good design must dress for the occasion.

A well-designed visual is a potent and valuable tool that has the potential to change perceptions and evoke certain emotions in an audience. And to do that, it must resonate with the viewer. In terms of data visualizations, their power lies in translating lengthy sets of data concisely and intentionally into an easy-to-understand story, making our lives more informed and better able to make decisions. Since humans aren't purely logical beings, but emotionally driven decision-makers, understanding the science behind how to connect with your audience emotionally will drive your designs to their memories.

Any visual design, from road signs to a data viz, involves a good knowledge of the elements and principles of design. In pretty much every visual, you will find design elements such as lines, shape, type, color, etc. Choosing the best placement, orientation, and grouping of these components is a skill that comes through consistent practice. Having an in-depth knowledge of the elements and principles of design will enable you to develop confidence and create informative visuals for your audience.

Mastering your understanding of layout and composition principles will bring it all together for your audience. A well-designed layout can be used as a template or style guide to work from for future projects, saving you time and energy. Using techniques such as a grid to organize, focusing on balance, and creating a visual hierarchy can build a solid home for your data story. A compelling data story consists of three cohesive parts: introduction, key message, and conclusion. While it is enjoyable to dive into the aesthetics of your design, you must not sacrifice function at the same time. Don't forget to GRAB, GUIDE, PING. That way, you can make your charts pretty AND smart.

Designing a visual is no easy feat (clearly), but it can be made effortless by investing your time in learning the best steps to follow while putting it together. This book is here to encourage, guide, and point you towards a better understanding of visual communication design that will work alongside your data science knowledge to enable you to design with a creative and sensitive approach. Consider your audience. Empathize with their disabilities, challenges, needs, wants, and most of all, emotions. This knowledge is as vital for a designer as elements of design. Anyone with enough practice and experience can design a good visual. Still, it takes perception, consistency, and empathy to create an excellent visual that resonates with your audience and remains long after viewing it.

To conclude, the true goal of creativity and design is to express your ideas, stories, and thoughts in the best way possible. What one person might consider their favorite method may differ from another, and so on, which is so exciting about the possibility for your future work. There is unlimited potential for you now. It is up to you to discover and spend time with the creative half of your brain and realize new skills. Whatever inspiration, motivation, or knowledge you have picked up from this book, take the next step and apply it to your work. Make it yours. A design is a bridge between the designer and the viewer. It creates a connection and conversation that didn't exist before. Let's get chatting.

YOU DID IT!

Hi there.

If you've made it this far, you have finished reading this book. I just wanted to give my sincerest thanks and appreciation for your time, and I hope that you found value in the chapters and inspiration between the lines.

If you'd like to be an inspiration for someone else and help them take a step toward realizing their potential, it just takes a small act of paying it forward.

To make that happen, all you have to do (this can take less than 60 seconds) is leave a review. If this book has been of any value so far, would you kindly take a brief moment and leave an honest review? I read every one and literally thank each person out loud.

To leave a review, please scan the QR code below

ACKNOWLEDGMENTS

I would like to first and foremost give thanks to my family. Creating a book isn't an easy task, and it was certainly a learning experience. I am grateful for my family's patience, confidence, and encouragement during this process. Mom, Dad, Lauren, Corwin, I love you guys. I would also like to thank my little tribe of fellow bookmakers. I am always grateful to have you to lean on when it feels like a marathon that never ends. I hope our group stays together and cheers each other on for many years to come. Jake, LT, Valerie, Tyler, Vince, Kimberly, Billy, Carleshia let's keep moving forward! I would also like to thank my dear friend Andrew M. for being my DesignEar on this project. Your advice and guidance is always appreciated. Big thanks to Erin T., who kindly lent her eyes to double-check things behind me. And a massive thank you to Dan L. for being a helpful mentor and guide in the early days of making this book. Thank you thank you to everyone who lent an ear, an eye, or a shoulder. I'll probably borrow them again if you don't mind.

BIBLIOGRAPHY

A. (2022, May 2). *Jewelry! Why Do Raccoons Like Shiny Things?* In To Yard. https://www.intoyard.com/why-raccoons-like-shiny-things/

Adobe Express. (n.d.). *8 basic design principles to help you make awesome graphics.* Adobe. https://www.adobe.-com/express/learn/blog/8-basic-design-principles-to-help-you-create-better-graphics

Andrews, R. J. (2019). *Info we trust: How to inspire the world with data*. John Wiley & Sons.

Apple Podcasts. (2019a, April 12). *EP#3 - Joshua Smith.* https://podcasts.apple.com/us/podcast/ep-3-joshua-smith/id1457065616?i=1000434827031

Apple Podcasts. (2019b, December 10). *EP#18 - Cole Knaflic.* https://podcasts.apple.com/us/podcast/ep-18-cole-knaflic/id1457065616?i=1000459215108

Balliett, A. (2020). *Killer Visual Strategies: Engage Any Audience, Improve Comprehension, and Get Amazing Results Using Visual Communication*. John Wiley & Sons.

BetterHelp Editorial Team. (2021, February 1). *The Visualization Definition And How It Transforms Your Life | BetterHelp.* Better Help. https://www.betterhelp.com/advice/visualization/the-visualization-definition-and-how-it-transforms-your-life/

Bremer, N., & Wu, S. (2021). *Data Sketches: A journey of imagination, exploration, and beautiful data visualizations*. CRC Press.

Carton, A. (2020, April 21). *What Is Typography? Why Is It Important For Graphic Designers?* Designhill. https://www.designhill.com/design-blog/what-is-typography-why-is-it-important-for-graphic-designers/

Cesal, A. (2021, October 28). *What Are Data Visualization Style Guidelines?* Nightingale. https://nightingaledvs.com/what-are-data-visualization-style-guidelines/

Clear, J. (2018). *Atomic Habits Tiny Changes, Remarkable Results*. Priyansh Rawal.

Dabner, D., Stewart, S., & Vickress, A. (2017). *Graphic design school: the principles and practice of graphic design*. John Wiley & Sons.

Darstaru, A. (2022, March 17). *Elements and Principles of Design To Use For Inspiration in 2021*. Creatopy Blog. https://www.creatopy.com/blog/elements-and-principles-of-design/

DeLong, S. (2021, July 1). *How to Create a Data Visualization Style Guide to Tell Great Stories (Part 2)*. Depict Data Studio. https://depictdatastudio.com/how-to-create-a-data-visualization-style-guide-to-tell-great-stories/

Design Principles: Visual Perception And The Principles Of Gestalt. (2014, March 29). Smashing Magazine. https://www.smashingmagazine.com/2014/03/design-principles-visual-perception-and-the-principles-of-gestalt/

Dykes, B. (2019). *Effective data storytelling: how to drive change with data, narrative and visuals*. John Wiley & Sons.

Emery, A. K. (2022, April 26). *The Data Visualization Design Process: A Step-by-Step Guide for Beginners*. Depict Data Studio. https://depictdatastudio.com/data-visualization-design-process-step-by-step-guide-for-beginners/

Few, S. (2004). Show me the numbers. Analytics Pres.

Gilbert, R. M. (2019). *Inclusive design for a digital world: Designing with accessibility in mind*. Apress.

Gunn, E. (2021, June 8). *How to Use Whitespace, the Punctuation Between Visual Elements*. Nightingale. https://nightingaledvs.com/how-to-use-whitespace-the-punctuation-between-visual-elements/

Hadjiosif, S. (2021a, November 7). *Why Is Design So Important In Our Society?* Terra Movement | Artivism. https://www.terramovement.com/why-is-design-important-in-our-society/

Hadjiosif, S. (2021b, November 7). *Why Is Design So Important In Our Society?* Terra Movement | Artivism. https://www.terramovement.com/why-is-design-important-in-our-society/

Hossain, P. (2018, June 1). *Preattentive Processing and How to Use it in Design*. Medium. https://uxplanet.org/preattentive-processing-and-design-e59eba74373e

Johnson, A. (2019, April 10). *EYE MOVEMENT DURING READING*. LinkedIn. https://www.linkedin.com/pulse/eye-movement-during-reading-andrew-johnson/

Kalluri, K. (2022, April 21). *Importance, Purpose, and Benefit of Data Visualization Tools!* Business Analytics Platform with

Pre-Built Insights | SplashBI. https://splashbi.com/importance-purpose-benefit-of-data-visualization-tools/

Kamps, H. J. (2022, April 14). *What's the difference between a camera and a human eye?* Medium. https://medium.com/photography-secrets/whats-the-difference-between-a-camera-and-a-human-eye-a006a795b09f

Khoo, A. (2021, June 8). *Why We Find Joy and Value in Creating Data Visualizations*. Nightingale. https://nightingaledvs.com/why-we-find-joy-and-value-in-creating-data-visualization/

Kirk, A. (2016). *Data visualisation: A handbook for data driven design*. Sage.

Kirk, A. (2021, June 8). *What Makes A Data Visualisation Elegant?* Nightingale. https://nightingaledvs.com/what-makes-a-data-visualisation-elegant/

Kleon, A. (2012). *Steal like an artist: 10 things nobody told you about being creative*. Workman Publishing.

Krum, R. (2013). *Cool infographics: Effective communication with data visualization and design*. John Wiley & Sons.

Malamed, C. (2009). *Visual language for designers: Principles for creating graphics that people understand*. Rockport Publishers.

Making Internet Things, Part 2: Design. (n.d.). The Pudding. https://pudding.cool/process/how-to-make-dope-shit-part-2/

Mbazzi, F. B., Nalugya, R., Kawesa, E., Nimusiima, C., King, R., Van Hove, G., & Seeley, J. (2020). The impact of COVID-19 measures on children with disabilities and their families in Uganda. *Disability & Society*, 1-24.

McCoy, E. (2020, January 6). *The Difference Between Infographics and Data Visualization*. Killer Visual Strategies. https://killervisualstrategies.com/blog/difference-between-infographics-data-visualization.html

Mcleod, S. (n.d.-a). *Long-Term Memory | Simply Psychology*. Simply Psychology. https://www.simplypsychology.org/long-term-memory.html

Mcleod, S. (n.d.-b). *Multi Store Memory Model | Simply Psychology*. Simply Psychology. https://www.simplypsychology.org/multi-store.html

Mcleod, S. (n.d.-c). *Short-Term Memory | Facts, Types, Duration & Capacity - Simply Psychology*. Simply Psychology. https://www.simplypsychology.org/short-term-memory.html

Müller-Brockman, J. (2010). *Grid systems in graphic design: Raster syteme für die visuelle gestaltung ein Handbuch für Grafiker, Typografen und Ausstellungsgestalter: a visual communication manual for graphic designers, typographers and three dimensional designers*. Niggli.

Muth, L. C. (2021, July 2). *What to consider when visualizing data for colorblind readers*. Datawrapper Blog. https://blog.-datawrapper.de/colorblindness-part2/

Nordlinder, E. (2022, February 26). *Designing with your brain in mind — Using Pre-attentive Attributes*. Medium. https://elias-nordlinder.medium.com/designing-with-your-brain-in-mind-using-pre-attentive-attributes-977a7af54b58

Owen, D. (2021, December 12). *Designing with memory in mind - UX Collective*. Medium. https://uxdesign.cc/designing-with-memory-in-mind-f109c15f12ff

Perera, A. (2021, February 1). *Sensory Memory | Simply Psychology*. Simply Psychology. https://www.simplypsychology.org/sensory-memory.html

The Personality Layer. (2012, July 18). Smashing Magazine. https://www.smashingmagazine.com/2012/07/the-personality-layer/

Philips, M. (2018, March 13). *Design Psychology and the Neuroscience of Awesome UX*. Toptal Design Blog. https://www.-toptal.com/designers/ux/design-psychology-neuroscience-of-ux

Principles Designs. (n.d.). Getty. https://www.getty.edu/education/teachers/building_lessons/principles_design.pdf

R. Heath, D. (2022). *Storytelling with data: a data visualization guide for business professionals*: by Cole Nussbaumer Knaflic, John Wiley & Sons, 2015, 267 pp., 19.25(paperback), 24.00 (Kindle Edition), ISBN 978-1119002253.

Quibell, D. A., Selig, J. L., & Slattery, D. P. (2019). *Deep Creativity: Seven Ways to Spark Your Creative Spirit*. Shambhala.

Schwabish, J. (2021). *Better data visualizations: A guide for scholars, researchers, and Wonks*. Columbia University Press.

Schwabish, J. (2021, April 20). *Why Your Organization Needs a Data Visualization Style Guide*. PolicyViz. https://policyviz.-com/2021/03/16/why-your-organization-needs-a-data-visualization-style-guide/

Steinbach, K. (2018). *Creative practices for visual artists: Time, space, process*. Routledge.

Stevens, R. (2020). *Powered by Design*. Rocky Nook, Inc..

Tomboc, K. (2020, September 24). *What is Information Design and Why It Matters Now More Than Ever.* Simple Infographic Maker Tool by Easelly. https://www.easel.ly/blog/what-is-information-design/

Tomboc, K. (2021a, May 19). *Data Visualization Guide: Choosing the Right Chart to Visualize Your Data*. Simple Infographic Maker Tool by Easelly. https://www.easel.ly/blog/types-of-graphs-and-charts-for-visualizing-data/

Tomboc, K. (2021b, May 19). *Data Visualization Guide: Choosing the Right Chart to Visualize Your Data*. Simple Infographic Maker Tool by Easelly. https://www.easel.ly/blog/types-of-graphs-and-charts-for-visualizing-data/

Tufte, E. R., Goeler, N. H., & Benson, R. (1990). *Envisioning information* (Vol. 126). Cheshire, CT: Graphics press.

Triberti, S. (2017). *Developing Emotional Design: Emotions as Cognitive Processes and their Role in the Design of Interactive Technologies*. Frontiers. https://www.frontiersin.org/articles/10.3389/fpsyg.2017.01773/full

Valuing The Eyes | Nikon Corporation Healthcare Business Unit. (n.d.). Healthcare Nikon. https://www.healthcare.nikon.com/en/eyecare/valuing-eyes/

Velarde, O. (2021a, March 19). *What Is Information Design and Why Is It Important?* Visme Blog. https://visme.co/blog/information-design/

Velarde, O. (2021b, March 19). *What Is Information Design and Why Is It Important?* Visme Blog. https://visme.co/blog/information-design/

Walter, A. (2011). *Designing for emotion* (Vol. 9). New York: A book apart.

Whalen, J. (2019). *Design for How People Think: Using Brain Science to Build Better Products*. O'Reilly Media.

What is Information Design? | SEGD. (n.d.). Segd. https://segd.org/what-information-design

Why Is Layout Important in Graphic Design? (n.d.). Flux Academy. https://www.flux-academy.com/blog/why-is-layout-important-in-graphic-design

Yau, N. (2011). *Visualize this: the FlowingData guide to design, visualization, and statistics*. John Wiley & Sons.

Your Company for Infographic Design & More. (2020, May 29). Killer Visual Strategies. https://killervisualstrategies.com/infographics

Made in the USA
Monee, IL
14 October 2023

44597626R00118